Praise for The Digital War

"The Fourth Industrial Revolution may impose significant challenges for many of the smaller countries, especially those that are lack of technology resources and have large labor forces that might be replaced by AI. But it may also be a leapfrogging opportunity for them, if those countries embrace the new technology revolution just as keenly and the Digital Silk Road from China provides necessary support. Winston Ma's latest book on digital economy is a cutting-edge text that spotlights the digital transformation in China and its global impact through the Silk Road in the cyberspace."

– Professor Klaus Schwab, Founder and Executive Chairman of the World Economic Forum

"As the United States and China head down the ill-conceived path of digital and technology decoupling, The Digital War must be read before these policies become irreversible. Winston's Ma's insider's view of China's digital development takes the reader down a path too rarely tread by those trying to understand China."

– Stephen A. Orlins, President, National Committee on U.S.–China Relations

"If you're not paying attention to China's digital transformation, you're missing the emergence of the world's new technology superpower, one whose innovations will drive rapid economic advances and alter the balance of global economic power. Thankfully, Winston Ma has delivered exactly the kind of comprehensive, timely and thought-provoking account you need to get up to date. Don't miss the vitally important read."

– Michael J. Casey, co-author, *The Truth Machine: The Blockchain and the Future of Everything*

"China is determined to make innovation an engine for the next stage of the country's development, and no sector has been more creative or dynamic than the mobile economy, which in some areas has surpassed even the United States. Winston Ma's deep dive into this fiercely competitive, constantly evolving industry dissects the companies, personalities and forces that are transforming China and that inevitably will influence commerce far beyond its shores."

– John L. Thornton, Co-Chairman, Brookings Institution

"Having worked at both the Wall Street and China Investment Corporation, Winston Ma seems eminently qualified to give both Western and Chinese readers a unique perspective on the development of China's digital economy. This is the go-to-guide for companies and investors seeking to understand China's grand tech ambitions, what the Chinese players' strategy is, and where the huge opportunities are."

– Dr. Chen Datong, Founding Partner, West Summit Capital

"The book is appropriately titled, The Digital War, but my great hope is that we can use technology to build a greater understanding between East and West."

– Anthony Scaramucci, Founder & Managing Partner of SkyBridge

Praise for
CHINA'S MOBILE ECONOMY
(*the beginning of Winston Ma's pentalogy on China's Digital Revolution and Tech Power*)

"One of the best 2016 business books for CIOs".

– i-CIO.com

"Every company – and every country – must succeed at digitization to compete successfully in the twenty-first century. Winston Ma delivers the rare book that is both an outstanding survey of a fast-changing and vitally important economic landscape and a delightful 'field guide' that will enrich your understanding of what's really happening on the ground".

– Dominic Barton, Global Managing Partner, McKinsey & Company

"Winston has written a first of its kind – timely, insightful, and eminently readable analyses of the world's fastest growing mobile economy. A must read for anyone interested in China, the mobile economy, or technology more broadly. Eye opening and thoroughly enjoyable".

– Reuben Jeffery III, President and CEO, Rockefeller & Co., Inc

"As the world moves to mobile technologies, and with China now the world's largest market of Internet users, all stakeholders have to think about China's economy, market, and society from a completely new perspective. This is an indispensable book for understanding the emerging shape and scale of opportunities in the Middle Kingdom and beyond".

– Rod Beckstrom, Co-author, *The Starfish and the Spider: The Unstoppable Power of Leaderless Organizations* and former President and CEO of ICANN

"Chinese society is experiencing a rapid transformation, becoming increasingly industrialized and digital-based. The Chinese internet population has officially entered into the age of mobile internet. This extraordinary book explains how the internet has been the engine that catapults commercial activities from offline to online and towards ubiquity".

– Dr. Xiaodong Lee, President and CEO, China Internet Network Information Center (CNNIC)

THE DIGITAL
WAR

WINSTON MA
FOREWORD BY ANTHONY SCARAMUCCI

THE DIGITAL
WAR

HOW CHINA'S TECH POWER SHAPES THE FUTURE OF AI, BLOCKCHAIN AND CYBERSPACE

WILEY

Registered Office(s)
John Wiley & Sons, Ltd., The Atrium, Southern Gate, Chichester, West Sussex, PO19 8SQ, United Kingdom

For details of our global editorial offices, customer services, and more information about Wiley products visit us at www.wiley.com.

Wiley also publishes its books in a variety of electronic formats and by print-on-demand. Some content that appears in standard print versions of this book may not be available in other formats.

Library of Congress Cataloging-in-Publication Data

9781119748915 (paperback); 9781119748892 (ePDF); 9781119748984 (epub); 9781119749073 (obook)

Cover Design: Wiley
Cover Image: © US America and China flags
on chess pawns © rawf8/Shutterstock,
Tik Tok icon © Icon Lab/Shutterstock

Set in size of 12/14 pt and ITC New Baskerville Std by SPi Global, Chennai
Printed and bound by CPI Group (UK) Ltd, Croydon, CR0 4YY

10 9 8 7 6 5 4 3 2 1

To Angela – I love you dearly

Contents

Foreword

The Digital War to End War

The explosion of China's mobile economy has paved the way for an even more transformational technology revolution in the country. So, naturally, Winston Ma is back with a tour-de-force on the latter subject, having enlightened us on the former only four short years ago. Ma, with his impressive business and academic background combined with a unique multicultural life experience, is the perfect messenger to explain the current and future state of the world's "digital war".

"If data is the new oil, then China is the new OPEC". While now cliché, the catchphrase provides a useful framework for understanding the progression of digital power. With a population of 1.4 billion people, a society of mobile-first internet users, and dominant super-apps in WeChat, Alibaba, and Meituan, China has a massive set of data from which to train AI systems – the more promising avenue for AI development (as opposed to rules-based AI) for the foreseeable future.

However, the availability of data is only one-half of the equation when it comes to technological advancement. The other half is the ability to collect and process that data dynamically. In addition to having the world's greatest reservoir of data, China has become the leader in developing tools to harness it. For example, it is the leader in developing 5G infrastructure. President Xi Jinping has publicly placed an emphasis on building platforms and providing incentives to incubate blockchain startups. Perhaps, most importantly, China has committed to supporting and nurturing the world's greatest tech minds, especially in fields like deep learning, at a time when America (at least temporarily) is attempting to wall itself off.

Ma begins the book with a discussion of the "Sputnik moment" for China when it comes to developing advanced technologies. When AlphaGo, an artificially intelligent system developed by Google's DeepMind lab, handily defeated world champion Ke Jie in a three-game series of the popular game "Go" in May 2017, it spurred an all-out commitment by the government and business leadership in the country to dominate the future of AI.

As a result of a laser focus on digital transformation, the portrayal of China as a copycat of intellectual property is now outdated. While China's tech community used to operate like a rules-based AI, mimicking American technology companies, it now operates more like a neural network, originating ideas and learning through data analysis. The most obvious example of this phenomenon is Chinese technology giant Meituan, which was originally launched based on the Groupon model in the United States. While Groupon ultimately failed, Meituan has flourished into one of China's most influential tech giants.

The success of China's campaign to become the dominant global player in data-driven technologies also leads the reader to a philosophical question about the role of government in society. Government intervention has been an accelerator, not a deterrent, to the development in China of the "iABCD" technologies—the Internet of Things, AI, blockchain, cloud computing, and data analytics. While US politics are today characterized by reactive short-termism and divisiveness, China is effectively executing on 20-, 50-, and 100-year plans. Just as I believe Chinese leaders could learn from elements of the American system, the United States must learn from the rapid progress made by China.

While Ma's earlier book, *China's Mobile Economy*, is a great prequel to *The Digital War*, he does a brilliant job of providing stand-alone context within his latest work. An American audience may only have learned about WeChat due to President Donald Trump's efforts to block the app in the United States. However, understanding the scope of WeChat, one of the most powerful apps in the world, is key to understanding the rise of China's digital prowess. The goofy short-video platform of TikTok, on the other hand, is a more recent phenomenon, and it has amassed huge number of US users since the COVID-19 pandemic. In fact, Bytedance, the parent company of TikTok, is the only technology firm, bar Apple, with more than 100 million users both in China and in the Western markets.

Ma has produced a terrific and highly accessible field guide to understanding how the digital economy is accelerating in China, and what that rapid growth means for China's place in the world. (Both WeChat and TikTok, e.g., are the subjects of US President Donald Trump's executive order to restrict their operations in the United States.) It is an honest and balanced account of the opportunities being seized on by the Chinese and the significant challenges that remain. While the innovation that gave rise to China's mobile economy was impressive, the possibilities created by its progress in the data economy are virtually limitless.

This book is appropriately titled *The Digital War*, but my great hope is that we can use technology to build a greater understanding between East and West. At a time when the United States is pulling away from the rest of the world under President Trump, China is filling the void. The Belt and Road and Digital Silk Road initiatives are strengthening infrastructure, trade, and investment links between China and emerging market economies. However, the internet is also bifurcating in unproductive ways. Every country in the world is now being forced to choose between China and the United States. In my opinion, the immense potential of Sino-American collaboration is too great an opportunity to miss.

While certain elements of our cultures will likely always remain incompatible, the United States and China have the ability to address myriad intractable global problems together, especially relating to health, climate, and security. The answer is not pulling out of the World Health Organization (WHO) or the Paris Climate Accord. The answer is using data and technology to produce better, fairer, supranational systems and more effective, innovative solutions. Smart, open-minded young experts like Ma – and those who heed the analysis contained in his book – are the ones who will make that possible.

Anthony Scaramucci
Founder and Managing Partner of SkyBridge

Author's Notes and Acknowledgments

Winston Ma

In the middle of 1990s, the early days of China's tech and Internet boom, I majored in electronic materials and semiconductor physics at Fudan University in Shanghai. Aiming for graduate studies in the United States, I diligently studied English for the TOEFL and GRE exams, and I also took a national exam for a professional certificate that is no longer relevant two decades later—"software programmer".

Back then, China had so few software programmers that the central government organized national qualification exams to encourage the young generation to study computer science. Sensing the tremendous potential of China's tech revolution, I sat in a one-day exam to solve coding problems in C, Fortune, and Pascal languages before I became a "nationally certified software programmer". Today, however, those programming languages are "old" for coding, and there is no need for such a national exam because every year hundreds of thousands of Chinese college students graduate from computer science majors—even highly specialized major like artificial intelligence (AI).

No doubt, over the past two decades, China has evolved to be a worldwide tech leader. Thus, in 2016, I shifted my focus to China's digital innovation for my book series on China with *China's Mobile Economy: Opportunities in the Largest and Fastest Information Consumption Boom* (also with Wiley publishing).

Compared to my previous book, *Investing in China: New Opportunities in a Transforming Stock Market*, published in 2006, the book on China's internet-based "mobile economy" was a much more challenging project. For one thing, China's economic model was transformed from decades-long double-digit rapid growth into a more sustainable growth model based on innovation and consumption. For another, the explosive growth of smartphone users, e-commerce, and online content consumption and creation led to a digital revolution in almost all industries and business sectors.

What I never expected then was that China's digital transformation would become so dynamic and powerful that I would have to write one book each year to cover its leap forward. After *China's Mobile Economy* (English, 2016), I published my book serial in different languages—*Digital Economy 2.0* (Chinese, 2017), *The Digital Silk Road* (German, 2018), and *China's AI Big Bang* (Japanese, 2019)—before I returned to work with Wiley again to publish this book, in English, with the China–US tech war in the background.

* * *

A book on such a complex and fast-moving topic would not have been possible if I had not been blessed to work with and learn from an amazing group of mentors in business, law, and tech investments.

My deepest thanks go to Dr. Rita and Gus Hauser at the New York University (NYU) School of Law, and John Sexton, the legendary Dean of NYU Law School when I was pursuing my LL.M degree in Comparative Law. My PE/VC investing, investment banking, and practicing attorney experiences all started with the generous Hauser scholarship in 1997.

During his decade-long tenure as the President of NYU, John kindly engaged me at his inaugural President's Global Council as he developed the world's first and only GNU (global network university). My NYU experience was the foundation for my future career as a global professional working in the cross-border business world. Great thanks to John Sexton, the legendary Dean of NYU Law School when I was pursuing my LL.M degree in Comparative Law.

My sincere appreciation to both Mr. Lou Jiwei and Dr. Gao Xi-qing, the inaugural Chairman and President of China

Investment Corporation (CIC), for recruiting me at its inception. One of the most gratifying aspects of being part of CIC is the opportunity to be exposed to a wide range of global financial markets' new developments. The unique platform has brought me to the movers and shakers everywhere in the world, including Silicon Valley projects that linked global tech innovation with the Chinese market.

The same thanks go to Chairman Ding Xue-dong and President Li Ke-ping, whom I reported to at CIC in recent years. Similarly, thanks to Linda Simpson, senior partner at the New York headquarters of Davis Polk & Wardwell, and Santosh Nabar, Managing Director at the New York headquarters of J. P. Morgan. Those two former bosses on Wall Street gave me a foundation to develop a career in the global capital markets.

Many thanks to Mr. Jing Liqun, President of Asian Infrastructure Investment Bank (AIIB) and formerly the Supervisory Chairman of CIC, for educating me about Shakespeare's works besides guiding me professionally. The readings on Hamlet, Macbeth, and King Lear improved my English writing skills, and hopefully, the writing style of the book serial is more interesting and engaging than my previous finance textbook *Investing in China.*

For such a dynamic book topic, I benefited from the best market intelligence from a distinctive group of institutional investors, tech entrepreneurs, and business leaders at the World Economic Forum (WEF), especially my peers at the Young Global Leaders (YGL) community and Council for Digital Economy and Society. Professor Klaus Schwab, Founder and Executive Chairman of the World Economic Forum, has a tremendous vision of a sustainable, shared digital future for the world, which is an important theme of this book.

My gratitude goes to many other outstanding friends, colleagues, practitioners, and academics who provided expert opinions, feedback, insights, and suggestions for improvement. For anecdotes, pointers, and constant reality checks, I turned to them because they were at the front line of industry and business practices. I would particularly like to thank the friends at the West Summit Fund, which I set up at CIC in 2009 for cross-border investments between Silicon Valley and China. Over the last 10 years, we've had great fun together.

On its journey from a collection of ideas and themes to a coherent book, the manuscript went through multiple iterations and a meticulous editorial and review process by the John Wiley team, led by the book commissioning editor Gemma Valler. Our long-term collaboration started with my 2016 book, and in 2020, just months before this book, we worked together and released another new work, *The Hunt for Unicorns: How Sovereign Funds Are Reshaping Investment in the Digital Economy* (October 2020). The managing editor Purvi Patel and copyeditor Barbara Long contributed substantially to the final shape of the book. Special thanks to Gladys Ganaden for her design of the book cover and figures.

And last in the lineup but first in my heart, I thank my wife, Angela Ju-hsin Pan, who gave me love and support. You are a true partner in helping me frame and create this work. Thank you for the patience you had while I wrecked our weekends and evenings working on this book.

About the Author

Winston Wenyan Ma, CFA & Esq.

Winston Ma is an investor, attorney, author, and adjunct professor in the global digital economy. He is one of a small number of native Chinese who have worked as investment professionals and practiced as capital markets attorneys in both the United States and China. Most recently, for 10 years, he was Managing Director and Head of North America Office for China Investment Corporation (CIC), China's sovereign wealth fund.

At CIC's inception in 2007, he was among the first group of overseas hires by CIC, where he was a founding member of both CIC's Private Equity Department and later the Special Investment Department for direct investing (Head of CIC North America office 2014–2015). He had leadership roles in global investments involving financial services, technology (TMT), energy, and natural resources sectors, including the setup of West Summit (Huashan) Capital, a cross-border growth capital fund in Silicon Valley, which was CIC's first overseas tech investment.

Prior to that, Ma served as the deputy head of equity capital markets at Barclays Capital, a vice president at J.P. Morgan investment banking, and a corporate lawyer at Davis Polk & Wardwell LLP. A nationally certified software programmer as early as 1994, Ma is the book author of *Investing in China: New Opportunities in a Transforming Stock Market* (Risk Books, 2006), *China's Mobile Economy: Opportunities in the Largest and Fastest Information Consumption Boom* (Wiley, 2016; among "best 2016 business books for CIOs"), *Digital Economy 2.0* (2017, Chinese),

The Digital Silk Road (2018, German), and *China's AI Big Bang* (2019, Japanese).

He is also an adjunct professor at NYU School of Law. Relating to his course "Sovereign Investments & Regulations", he published *The Hunt for Unicorns: How Sovereign Funds Are Reshaping Investment in the Digital Economy* (2020) with Wiley.

Ma has served on the boards of multinational listed and private companies. He was selected a 2013 Young Global Leader at the World Economic Forum (WEF) and has been a member of the Council for Long-Term Investing and Council for Digital Economy and Society. He is a member of New York University (NYU) President's Global Council since its inception, and in 2014, he received the NYU Distinguished Alumni Award.

The author can be reached on LinkedIn for comments and feedback on *The Digital War: How China's Tech Power Shapes the Future of AI, Blockchain and Cyberspace.*

Preface

In May 2017, China hosted the historical match of the *weiqi* game (the "Go" game) between Ke Jie, the world's No.1 ranked player and world champion, and the AI (artificial intelligence)-enabled computer Go program named AlphaGo, designed by the DeepMind Lab of the US internet giant Google. The Wuzhen showdown was ripe with suspense and symbolism—human versus machine, intuition versus algorithm, tradition versus modern, and with the AI machine's straight 3–0 win over the world's best human player, the unequivocal rise of the "digital economy".

Almost overnight, the internet business community in China started discussing "the second half" of the mobile economy era, which in 2013–2016 drove a boom in e-commerce and mobile entertainment. In particular, the image of a top human player crying at his loss to AI has triggered a great sense of determination and urgency among Chinese businesses and companies: either adapt to the fast-evolving technology of AI, big data analysis, and computer chips to upgrade – or be destroyed. Since 2017, the new key words have been *data* and *intelligence*.

As covered by my 2016 book *China's Mobile Economy* (published by Wiley and called one of "the best 2016 business books for Chief Information Officers" by i-CIO.com), China has built the world's largest mobile internet economy, powered by the world's largest group of internet users, smart devices, mobile applications, and social networks. China's digitally connected new middle class has led to a seismic change in the Chinese consumer market. Even for multinational corporations that have done business in China for many decades, a comprehensive rethink of their strategies in China may be necessary.

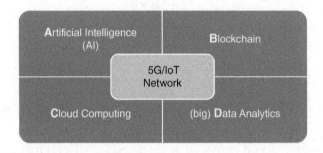

Today, China's <u>consumer-focused</u> Internet is transforming into a more <u>enterprise-oriented</u> Internet, characterized by faster 5G mobile networks and more advanced digital technologies, including the Internet of Things, artificial intelligence (AI), blockchain, cloud computing, and data analytics (iABCD). Across just about every industry sector, Chinese companies have rushed to learn how 5G iABCD digital technologies can be integrated into their businesses to unlock value from nontraditional angles (see figure above). In short, China is leaping forward from "mobile first" to "intelligence first".

For example, Tencent, one of the highest-valued Chinese internet companies, with its roots in gaming and online services, has made dramatic changes in response to the new trend. In 2018, for the first time in six years, Tencent announced a major restructuring to move from a consumer-only business toward one that caters to industries as well. The restructuring includes the creation of a new Cloud and Smart Industries Group, focusing on artificial intelligence, cloud service, Big Data and security; furthermore, Tencent has formed a new technology committee to better coordinate fundamental technology research across the company.

As a result, the next phase of development is referred to as the "second half game" of the mobile internet economy. The market expects far more profound and broader transformation of China's economy to come from "intelligent +" (data economy) than from earlier years of "internet +" (mobile economy). The years 2017 to 2018 were an even more important inflection point for China's digital transformation than the 2014–2015 inflection point (see the following figure). We are at the cusp of another major digital revolution in China.

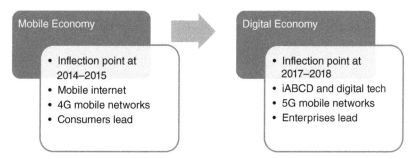

This shift has profound implications for emerging markets that are looking at China as a reference case when they work on their own digital transformation. That means they need to look beyond mobile phone and digital wallet; instead, they must start positioning themselves for the next phase of AI and the digital economy – now. The billion-user messaging service of the WeChat app (Tencent), the US$30 billion and more total GMV (Gross Merchandise Value) in 24 hours on the Single's Day, and the "smile-to-pay" functions creating a cashless society are already snapshots from the past.

For example, yes, the November 11 (Singles' Day) festival remains the world's largest online shopping day, beating Black Friday and Cyber Monday combined. However, the highlight of late is no longer the GMV number but the advanced data technologies involved. Because the logistics issue of Singles' Day—the inventory, distribution and delivery of numerous orders close to US$40 billion in a short span of time—is such a challenge, Alibaba's logistics affiliate, Cainiao, has used AI techniques and GIS (Geographic Information System) to determine the fastest and most cost-effective delivery routes in a variety of complex road networks, including both rural villages and crowded urban areas.

As such, China's leaping forward may also give the rest of the world (emerging markets) a sense of urgency. But it may also be a leapfrogging opportunity for them, if those countries embrace the new technology revolution just as keenly. From "mobile" to "digital", what's truly extraordinary is the decisive commitment from the Chinese government. It plans to become possibly the world's first AI superpower through its ambitious move to embrace the AI and Fourth Industrial Revolutions.

Furthermore, in late 2019 President Xi urged that China should accelerate the development of blockchain to "seize the opportunity", in remarks that marked the first major world leader to issue such a strong endorsement of the widely hyped—but still unproven—technology. During the 2020 coronavirus pandemic, China launched the "New Infrastructure" initiative in May, which acts as a stimulus to help stabilize the economy against COVID-19 disruptions, as well as a foundation for China to stay competitive and succeed in the age of Industry 4.0.

Meanwhile, the rise of China's tech power has also disrupted the global competition landscape, challenging the long-held dominance of America's Silicon Valley. In fact, the years of 2019–2020 may be remembered as an important inflection point for the global digital economy, from collaborative co-existence to head-on tension. It is critical for the United States and China to reach a new equilibrium to collectively lead innovation in the age of AI because at the core of the digital economy is the free flow of trade, capital, intellectual, and data. Global dialogue and cooperation are ever more important for a shared digital future of the world.

The book is organized in three parts, as follows.

Part One: From "Mobile Economy" to "Digital Economy"

The two chapters in Part One lay the foundation for the rest of book. China is the largest homogenous digital market in the world, unified by the same language, culture, and mobile payment system. They provide an overview of China's mobile internet revolution during the last few years and how the mobile economy set up the foundation for the forthcoming industrial revolution.

Chapter 1: China's Leap into 5G iABCD

The opening chapter is an overview of China's digital transformation. Chinese internet companies have turned themselves into tech companies, and they subsequently help companies from all sectors of the economy to digitally transform. Blockchain is the latest addition to the innovation mix, and the Blockchain BaaS services from tech giants are enabling entrepreneurs to focus on building new applications, leading to a wave of blockchain startups in China.

Chapter 2: The World's Largest Mobile Economy

Chapter 2 explains the evolution of mobile internet businesses in China during the past years. In addition to the unrivalled internet user population size, what also makes the China market unique is the fact that China is the largest "mobile first" and "mobile only" market in the world. One extraordinary example is Singles' Day (a Chinese holiday on November 11, now the world's largest shopping day), which in 2019 had more than US$38 billion GMV in 24 hours, connecting buyers and sellers from more than 200 countries across the world—with more than 90% transacted from mobile terminals.

Still, China's mobile internet market shows no sign of slowing down in innovation, and the competition between the leading giants and the hungry upcomers is ever fierce. The mobile infrastructure, corresponding data accumulation, and industries digitalizing their businesses along the way, collectively provide a powerful foundation for China's AI innovation and digital transformation today.

Part Two: China's Digital Transformation and Innovation

The chapters in Part Two cover in detail Chinese companies' digital transformation, which is reshaping retail, financial, entertainment, transportation, lifestyle services, education, smart hardware, and many traditional industries. China is one of the most interesting innovation centers in the world and where the new generation startups are pioneering 5G iABCD innovation.

Chapter 3: Big Data on the Digital Middle Class

There is no longer a single, one-size-fits-all definition of the Chinese consumer. The key to all industry sectors is to personalize products and services to serve the varying needs and demands of millions of Chinese consumers. In this chapter, the case studies of Hema "new retail", Pinduoduo C2M social e-commerce, and "Big Data" movies are all successful examples of consumer data-driven OMO (online-merge-offline) models. E-commerce giants such as Alibaba and JD.com had acquired "bricks-and-mortar" stores way before Amazon had acquired Whole Foods. Now they set up fully "digitalized" stores to integrate online and offline operations and data.

Chapter 4: The AI-powered Internet Celebrities and Fans Economy

This chapter covers the digital entertainment revolution in China, where average users, or fans, are at the center of the fans economy. New social media platforms are pivoting from "elite"-centered community toward the untapped grassroots crowds derived from smaller cities, younger age, lower social status, and less glamorous-looking groups. Numerous examples of the new fans economy are examined: the bullet screen of Bilibili, the boom of online novels, the revival of Weibo (China's Twitter), the struggles of video streaming sites like Youku and iQiyi, and the rise of short-form video platforms like Tik Tok and Kuaishou.

Chapter 5: The Heartland of Blockchain and Fintech

This chapter discusses many successful examples of China's internet finance and fintech innovation, including mobile payments, crowdfunding, consumer credit system, asset management, and corporate finance. China is far ahead of the rest of the world in terms of how widely internet finance is used. China's fintech success is not just from unprecedented technological innovation, but also from integrating tech tools with real-economy needs, which is best illustrated by its distinctive approach to Bitcoin and blockchain technology (supply chain finance is the best example of the latter).

Chapter 6: O2O and the Shared Economy

The shared bikes and then "everything share" (essentially "mobile renting") is a unique Chinese phenomenon. The common denominator of all the "sharing" ventures, from cars and bikes to basketballs and refrigerators to clothes and massage chairs to luxury handbags and phone chargers, is the mobile payment mechanism for which China is by far the global leader. Unlike the United States, mobile renting of everything is a bigger story—for China and emerging markets—than putting existing under-utilized cars (Uber) and spaces (Airbnb) to more frequent use.

The bigger "shared economy" is on the major O2O platforms like Meituan, Alibaba, and Didi. They create affordable infrastructure for startups by allowing them to share the on-demand transportation system and a massive fleet of delivery staff, as well as additional operational systems like smart payment and location-based targeted consumer marketing, rather than own them. When everyone was ordering take-outs during the coronavirus outbreak, Meituan and Ele.me cultivated a clan of "virtual restaurants" that operate only out of a kitchen.

Chapter 7: From C2C to 2CC: "Innovated in China"

Compared to their US counterparts, Chinese mobile apps are more advanced for content, social and commerce, and they are more transaction-based. The examples in this chapter show that the new trend is "to China Copy" (2CC), that is, Western companies copying new business models that are coming out of China. The new generation Chinese companies are shifting their focus, from "innovative models" to "innovative technology (iABCD)", from "online platform building" into "digital technology research". From copiers to originators, Chinese tech companies are showing the way forward with leading-edge advances that rival the West.

Chapter 8: Land of Big Data and Its Legal Framework

If data is the new oil, then China is the new OPEC. The country's massive population—nearing 1.4 billion people—offers researchers and start-ups valuable human data for AI training. Meanwhile, Chinese users are demanding more data privacy. After years of Chinese internet companies building business models around Chinese people's lack of awareness about privacy, users are becoming more knowledgeable, and they are getting angry about companies abusing their personal information. China is putting together a comprehensive legal framework to protect individual data rights from large internet platforms, and China's legal development as part of the global digital governance system simply cannot be ignored.

Part Three: Shared Digital Future

China in Asia and the United States in the West are forming two leading innovation centers of different strengths. China's digital economy is expanding overseas, both forming development partnerships with emerging markets and challenging America's tech supremacy.

Chapter 9: The Tech Cold War

China's ambitions and progress have led to talk of an artificial intelligence arms race with the United States. Cross-border tech investments are increasingly viewed through national security lenses by the two countries. In recent years, more and more deals have been blocked, corporations sanctioned, and AI exchanges restricted. Hence, the digital economy is in a vital conflict and crisis: the global tech world, together with at least part of the world economy, has now fractured into two – and potentially more, considering Europe, Japan, and other regions – spheres of influence. Nations and companies around the world are being forced to choose sides in a conflict that is fracturing global supply chains and tech innovation.

Chapter 10: The Silk Road in Cyberspace

"Digital Silk Road (DSR)" is the digital prong of China's Belt and Road Initiative (BRI), which seeks to strengthen infrastructure, trade, and investment links between China and the rest of the world. While many Western observers emphasize the data security risk of the DSR, the DSR initiative does help the unmet needs of digital connectivity in emerging markets, from online education in Africa, smart health in the Middle East, to local entrepreneurs in ASEAN countries. The DSR – at least the global connectivity concept behind it – is an important initiative for an increasingly fractured tech world. All global stakeholders must cooperate to ensure that digital innovation continues to serve as the growth engine for the world.

From "Mobile Economy" to "Digital Economy"

China is the largest homogenous digital market in the world, unified by the same language, culture, and mobile payment system. The first two chapters provide an overview of China's mobile internet revolution during the last few years and how the mobile economy set up the foundation for the subsequent industrial revolution.

China's Leap into 5G iABCD

- Blockchain with Chinese Characteristics
- AlphaGo, the Historical Match
- AI First—"Second Half Game" of Mobile Internet
- 5G iABCD New Infrastructure
- Digital Transformation in the Cloud
- BaaS Startup Innovation Ecosystem
- Splinternet and the Digital Silk Road

Blockchain with Chinese Characteristics

Since the coronavirus became a pandemic and people dived into remote working in March 2020, video-conferencing services have become vital components of this new way of life. Zoom, an until then little-known online platform, has emerged as the go-to service for not only virtual meetings and classroom lessons, but also church services, costume parties, romantic dates, and even wedding ceremonies. Zoom has quickly become the No. 1 video-conferencing platform in the United States, more popular than similar offerings from Google, Microsoft, and Facebook. Between December 2019 and April 2000, Zoom's daily meeting participants jumped 30-fold to 300 million in a mere five months.

Whereas Zoom is headquartered in the United States and listed on the Nasdaq exchange, the actual Zoom app appears to have been developed by companies in China. In its Securities and Exchange Commission (SEC) filings, Zoom has at least 700 employees who work in research and development out of its affiliated companies in China. The company founder is a native Chinese who had working experience at the US telecom giant Cisco. When Western corporations flocked to Zoom it could well have been their first time to use Chinese enterprise-solution software in their businesses. In the past, the US and European markets mostly learned about China's booming digital economy from the billion users of WeChat (the WhatsApp of China), mind-boggling e-commerce volume of Alibaba, or the goofy videos of TikTok.

Most of them, however, probably have missed a much bigger technology breakthrough in China in the same month, which further illustrates China's digital economy prowess—in the form of "hard tech" innovation—way more than pure mobile applications connecting online users. After more than six years of preparation, in April 2020 People's Bank of China (PBOC, the central bank) unveiled to the public its new digital currency, the world's first central bank digital currency (CBDC), known as the Digital Currency/Electronic Payment (DCEP), which could replace cash with a blockchain (like)-based solution.

Most likely, DCEP makes China the first major economy to adopt a native digital currency. DCEP is designed to function as the digitalization of physical cash (i.e. paper cash, coins, and banknotes) or just as the substitution of the base money supply (M0)—at least for now. Progress on the DCEP was broadly viewed as a reaction to Facebook's announcement in 2019 that it intended to launch *Lybra,* Facebook's planned blockchain-based digital currency, which is still trying to win approval from regulators. Unlike Bitcoin and other cryptocurrencies, built on the excitement regarding "decentralization", DCEP is run from a centralized database; nevertheless, DCEP is built with block-chain and cryptography, and it has incorporated blockchain's

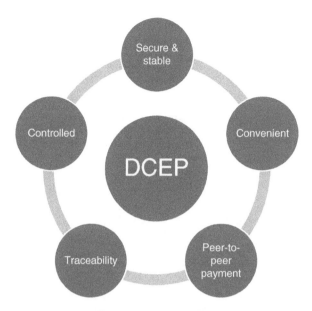

Figure 1.1   How is DCEP designed?
Data Source: Media Reports, April 2020.

key concepts, such as peer-to-peer payment, traceability, and tamper-proof-ness (see Figure 1.1).

Since its April launch, the new digital currency has been piloted in four major cities, Shenzhen, Suzhou, Chengdu, and Xiong'an New Area, the city to replace Beijing as the country's capital. (Leveraging the blockchain technology's strength in data management, the whole city of Xiong'an has a blockchain DNA. **See the "Vice Capital Built on Blockchain" box**.)

In the pilot zones, DCEP has been formally adopted into the cities' monetary systems, with some government employees having started to receive their salaries in the digital currency in May 2020. People can create a DCEP wallet in their commercial bank's mobile app and use the national digital currency for expenses like transportation, education, healthcare, and other consumer goods and services. Starbucks, McDonald's, and Subway chains in China, for example, were named on the central bank list of firms to test DCEP. In July, Chinese ride-hailing giant

Didi Chuxing entered a strategic partnership with the PBOC to test DCEP on its transport platform of over 500 million users. In August, China expanded DCEP trials to Beijing and several major cities—getting even closer to its official launch.

Vice Capital Built on Blockchain

Billed as "a strategy crucial for the millennium to come," the Xiong'an New Area is a new area of "national significance" launched by China on April 1, 2017, following the Shenzhen Special Economic Zone in the 1980s and Shanghai Pudong New Area in the 1990s. Designed for taking over "non-capital functions" from the hyper-crowding of Beijing, Xiong'an (meaning "brave and peace"), located about 100 km southwest of Beijing, will be a new home—a "vice capital"—for Beijing's colleges, hospitals, financial institutions, and state-owned enterprises' headquarters.

As China's "city of the future", Xiong'an has been designed to become a smart city zone for innovation. In this context, the integration and application of the Internet of Things (IoT) technologies, artificial intelligence, Big Data, and cloud computing has been strongly encouraged in the city planning. In particular, the new city is set to become the blockchain hub in China and has taken a lead in rolling out blockchain-based services.

The blockchain DNA of the megacity had started even before its construction. To build the vice capital over swampland, the government purchased land from local farmers. Those transactions are made via blockchain to keep them transparent and organized. The so-called Blockchain Land Compensation Distribution Platform distributes financial subsidies for relocated residents in the area. Since April 2020, DCEP, the blockchain-based digital currency, has been tested at hotels and convenience stores in the New Area.

In order to create a "beautiful forest to last a thousand years", every tree in Xiong'an will have its unique identification after being planted. The Xiong'an project team has built a system powered by blockchain, Big Data, and cloud computing technology to track the whole process for quality control, including seedling selection, excavation, packaging unloading, planting, and irrigation. Based on this system, each tree has a digital ID. After being encoded, all of its information, including its species, place of origin, planting location, and growth status, can be found in the database.

In parallel, as a model of "transparent Xiong'an", a blockchain-powered funding management platform is used to monitor the flow of cash and ensure the funds will be used exclusively for the digital forest project. The blockchain platform also tracks each worker's activities and directly pays salaries to their bank accounts. By the end of 2019, Xiongan had planted 14 million trees—on the blockchain.

China's DCEP has set off a global debate for its future potential to challenge the US dollar for primacy. To some, DCEP—the world's first central bank digital currency (CBDC) for a major country—may provide a functional alternative to the dollar settlement system that reigns supreme in the global economy for decades. For example, China could use DCEP to manage funding and transactions for its Belt and Road Initiative (BRI) of overseas infrastructure investments, extending its monetary sphere of influence.

This national digital currency push does not come as a surprise. In an October 2019 speech, Chinese President Xi Jinping declared blockchain "an important breakthrough," that would play "an important role in the next round of technological innovation and industrial transformation." President Xi urged that China should accelerate the development of blockchain to "seize the opportunity", in remarks that marked the first major world leader to issue such a strong endorsement of the widely hyped—but still unproven—distributed ledger technology (DLT). (By contrast, most governments in the West have been far more cautious.)

Calling for blockchain to become a focus of national innovation, President Xi's speech detailed the ways the Chinese government would support blockchain research, development, and standardization. To global policymakers, China's blockchain push seems to serve two of its strategic goals: ending the hegemony of the dollar and reducing dependence on the United States for foundational technologies. The central government's support for blockchain projects in China puts any initiatives or standards that emerge on the front foot, potentially hastening the mainstream application and use of the technology.

To be clear, President Xi's speech and DCEP is entirely about blockchain, far from a vote of confidence in Bitcoin or other cryptocurrencies. Blockchain is a form of distributed ledger that creates an online database where every participant can share and synchronize information. The data, maintained in chained records called "blocks", is not owned by any single authority (hence, the notion of "decentralization"). Such a decentralized, cryptography-based network mechanism enables immutable data retention and secure transmission. As a matter of fact, Chinese entrepreneurs have been on the crest of the blockchain wave since Bitcoin first gained traction.

The concept first emerged in the 1990s, but did not become reality until 2008, when Satoshi Nakamoto—a pseudonym whose true identity still remains unknown—reportedly launched Bitcoin, a digital currency that used blockchain to realize and store its value. China's startup ecosystem had aggregated enormous experience in building digital blocks before the release of the Ethereum platform in 2015, which marked a milestone in the second-generation blockchain system that embodied "smart contracts". The blockchain field was revolutionized by the implementation of layered smart contracts in the Ethereum platform to create applications other than money exchanges.

Along the way, China has become home to some of the world's largest cryptocurrency "mining farms"—data centers hosting the high-powered computers where the so-called miners (companies or passionate individuals) compete against others in the blockchain network to solve complex math puzzles and earn new coins. According to CoinShare's estimate in December 2019, approximately two-thirds of global Bitcoin mining (65%) happens in China: Sichuan province alone produces more than half (54%) of global hashrate (the parameter for mining capacity). Moreover, China manufactures most of the world's mining equipment. Chinese companies, such as Bitmain, Canaan, and MicroBT, are among the world's biggest manufacturers of Bitcoin mining gear.

Furthermore, the Chinese exchanges for cryptocurrencies (Bitcoins and other tokens) used to lead the world in terms of volume. The "coin" rush was interrupted by Chinese authorities in September 2017, which effectively banned all Initial Coin Offering (ICO) activity within China as "unauthorized and illegal public fundraising" and "unauthorized public sales of securities". The Chinese government also made illegal all cryptocurrency exchanges within the country. As a result, the market saw cryptocurrency trading and other related activities in China moving abroad due to tightened regulation.

Although China has cracked down on cryptocurrencies, shutting down all domestic crypto exchanges and banning all ICOs, blockchain technology itself is recognized as a revolutionary development by the government. After all the noise, hype, and speculation died down, the blockchain technology reached its third phase and has been integrated with the real economy since 2018 (see Figure 1.2). Real companies have come to the field to focus on application of the technology in solving real business problems to create solid values.

For enterprises, blockchain is expected to apply broadly in the business world to revolutionize how data is managed and shared. Blockchain can offer an excellent solution to synchronize data, especially sensitive information, across companies, industries and geographical boundaries. It can facilitate a trusted network that enables multiple parties to exchange data, information, and assets directly, as in the case of supply chain finance (see the details in Chapter 5 relating to fintech discussions) in the manufacturing sector. The technology is used widely across a range of industries in China: for banking, financial services, public services, healthcare, logistics, and

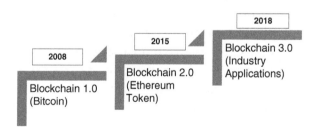

Figure 1.2   Blockchain from 1.0 to 3.0

smart manufacturing. If blockchain is mainstream anywhere, it's in China.

No doubt, the launch of DCEP and the ongoing rollout of blockchain applications should further support China's global technology ascent. Whereas blockchain technology is newly embraced by China as a frontier of innovation, the country is also making giant strides in advancing the 5G network, artificial intelligence (AI), and more emerging technologies—a digital rush started from a chess game in 2017.

AlphaGo, the Historical Match

In May 2017, the Chinese media was abuzz with reports about a historical match of the Go chess game (*weiqi*). It was a best-of-three match between Chinese player Ke Jie, the world's No.1 ranked player and world champion, and the AI-enabled computer Go program called AlphaGo, designed by the DeepMind Lab of the US internet giant Google. The match turned out to be one of China's most talked-about news events in 2017, attracting hundreds of millions of views on social media.

For many, the Wuzhen showdown was ripe with suspense and symbolism: uman versus machine; tradition versus modern; intuition versus algorithm; East versus West. Who would prevail? (**See the "Match Impossible" box.**)

In the end, a tearful Ke Jie became the hallmark image of this historic match. After losing 0–3 to AlphaGo, Ke Jie took off his glasses and wiped his eyes, the sound of his distress filled the room in which he had fought and lost. Subsequently, AlphaGo went on to defeat a team of five top Chinese professionals in a demonstration game in which they were allowed to lay down variations on a board and discuss among themselves, showing that even collective human minds couldn't beat the machine.

Match Impossible

China is the birthplace of *weiqi* (*Go* chess), an ancient board game played on a 19x19 grid. In *Go*, two players place black or white stones on the grid, each seeking to seal off the most territory. Historical records show it was played as early as the Zhou dynasty (1046 BC–256 BC). The match took place in Wuzhen, Zhejiang province, where there is a canal more than 1300 years old—a fitting venue for a game that dates back thousands of years. Wuzhen also hosted China's annual World Internet Conference, creating a parallel link to the digital power of AlphaGo.

In contrast to the long history of Go within Chinese culture, the development of AlphaGo was only three years old at the 2018 match. Go is seen as an extremely difficult game for computers to master because there are more possible board configurations in Go than there are atoms in the visible universe. Furthermore, human players believe that winning multiple battles across the board relies heavily on intuition and strategic thinking and that a software algorithm cannot simply memorize all combinations of board pieces, assess the situation by calculating all possible moves, and select a strategy to win, like in chess.

As such, Go has been a benchmark for measuring the human mind against artificial intelligence after IBM's Deep Blue beat chess grand master Garry Kasparov in 1997. For many years, there was little progress. More recently, the AlphaGo program developed by Google's DeepMind managed to analyze the game in a different way. AlphaGo used two sets of "deep neural networks" containing millions of connections similar to neurons in the brain—one that selects its next move while the other evaluates the decision.

The Google programmers provided AlphaGo with a database of 30 million board positions drawn from 160 000 real-life games to analyze, and the program was also partly self-taught, having played millions of games against itself following its initial programming ("machine learning"), all the while learning and improving. AlphaGo's success was considered the most significant yet for AI, due to the complexity of Go game, which has an incomputable number of possible scenarios, and in particular emphasizes the importance of "intuition" or "instinct" that is thought to be reserved for humans only.

After AlphaGo beat the best human player, Google developed a more advanced version, AlphaGo Zero, which was not trained by historical data of games between humans at all. Instead, AlphaGo Zero was only taught of the *Go* chess game rules before it started self-training by playing games against itself. Within a few days, AlphaGo Zero easily beat AlphaGo. Clearly, in addition to the quantity of data, other factors, like new algorithms, computing power, and the kinds of data available, may be just as valuable for AI training.

As if to respond to the humans' confusion, fear, and depression, at the end of the match the DeepMind Lab team announced that AlphaGo would retire from competing against human players. Instead, the team would largely shift toward using AI to solve problems in health, energy, and other fields. There was no doubt about AI's overwhelming superiority, at least in the game of Go.

First, the AI program showed more understanding of Go than humans, even to the extent of perfection. From time to time, AlphaGo put down seemingly randomly placed stones to set up winning positions. Those surprises kept coming in all three games, with the AI program making "unconventional" and "interesting" moves against Ke Jie. In a later interview, Ke Jie vowed never again to subject himself to the "horrible experience" because "he had had enough".

"For human beings", a visibly flummoxed Ke Jie commented with a resigned expression, "our understanding of the Go game is really very limited". Meanwhile, "AlphaGo to me is 100% perfection", he added, showing feelings of helplessness and depression. Even for the world's No.1 player, confronting an enemy that never makes mistakes and always picks the best possible moves ahead of its rival was no longer a competition, but torture.

Second, the AI program had no emotions or feelings, which seemed to be another advantage over humans. In close games, that may have given AlphaGo an edge. Toward the end of the second match, Ke Jie was visibly agitated, tugging his hair, rubbing his chest, and laying his head on the table from time to time. After the game he confessed that, when he thought he might have had a chance at winning in the middle of the game, he got too keyed up to keep calm. "I was very excited. I could feel my heart bumping", he said. "Maybe because I was so excited I made some stupid moves".

Third, the AI program made the Go game more interesting. One-time world champion Shi Yue commented that, in games between human players he had never seen moves like Alpha-Go's and was unlikely to in the future. The question to follow

Table 1.1   The speedy ascent (and retirement) of AlphaGo

November 2015	DeepMind organized a secret match with Fan Hui, Chinese 2-dan pro and winner of several European championships. AlphaGo won 3–2 in unofficial training games, and won 5–0 in the official match
January 2016	DeepMind published a paper in the journal Nature describing the AI system behind the AlphaGo version that beat Fan Hui. The team also announced a five-game match against Lee Sedol, the top player of the previous 10 years
March 2016	The upgraded version of AlphaGo played a best-of-five match against 9-dan Lee Sedol, the multiple world champion from South Korea, and won 4–1
January 2017	A new, upgraded version of AlphaGo (called "Master") won 60–0 against most top professionals from China, Korea and Japan in fast-move games (mostly 30 seconds per move)
May 2017	AlphaGo defeated 9-dan Ke Jie, the reigning top-ranked player from China, 3–0 in a best-of-three match
May 2017	DeepMind team announced that Alphago would "retire" from competing against human players

Note: For the *Go* chess, professional ranks in China, Japan, and Korea all start at 1-dan and go up to 9-dan, the best players being 9-dan.

is whether there is still value in human-versus-human games? If games between AI programs become more interesting and unpredictable, the existential value of professional Go players could be questioned.

Most strikingly, the leaps in AI power happened in a short period of time. The ascent of AlphaGo to the top of the Go world was distinct from the trajectory of machines playing chess games. Because of its vast number of possible scenarios (in the order of magnitude numbering 10 to the power of 360!), many professional players estimated that it may take AI at least another 10 years before it could outperform top human players. However, the development and perfection of AlphaGo spanned less than three years (see Table 1.1). "Last year, I think the way AlphaGo played [against Lee Sedol of South Korea] was still quite human-like, but today I think he plays like the God of Go", Ke Jie said after the game.

As such, AlphaGo's superior calculation power stripped Chinese audiences of their initial curiosity about AI and threw them into confusion. Almost overnight, the internet business community in China started discussing about "the second half game" of the mobile internet economy which, in 2013–2016, led a boom in e-commerce and online entertainment. Since 2017, the new key words have become *data* and *intelligence*, and the resolve to close the gap with—and quickly surpass—Silicon Valley in deploying AI is prevalent across the country.

AI First—"Second Half Game" of Mobile Internet

The image of the world's top player crying at his loss to AI has triggered a great sense of determination and urgency among Chinese businesses and companies about AI: either adapt the fast-evolving technology of AI, Big Data analysis, and computer chips (for AlphaGo, Google designed a special-purpose chip specifically for machine learning) to upgrade—or be destroyed.

As such, the largest internet companies, such as Alibaba (the e-commerce giant like Amazon) and Tencent (best known for its billion-user social messaging service WeChat) jumped on the AI wave to transform themselves into "intelligence first" companies. During the recent "mobile first" era (which feels like ages ago), their mobile platforms had accumulated vast amounts of user and transaction data; such data is now being leveraged, using AI, to solve practical operational challenges and drive new business models. (Hence, the notion of the "second half game".)

Tencent—the highest-valued Chinese internet company, which has roots in gaming and online services—made dramatic changes in response to the new trend. In 2018, for the first time in six years, Tencent announced a major restructuring to move from a consumer business toward one that caters for industry as well. The restructuring included the creation of a new Cloud and Smart Industries Group, focusing on AI, cloud services, Big Data and security; and another new group combining its social media, mobile internet, and online media operations for the purpose of strengthening internal coordination to compete with emerging competitors like Toutiao (the short video platform that owns the world-famous app TikTok).

Furthermore, Tencent formed a new technology committee that better coordinates fundamental technology research in different parts of the company. Tencent's restructuring marks a "significant strategic upgrade" for the company and comes as its main business of gaming and online services began to grow at a slower pace. As publicly stated by Pony Ma, Tencent's founder and chairman, "the next era of the internet is the industrial internet", and Tencent will, in reaction, develop new industry-facing services to "connect industries and consumers to build a more open ecosystem".

Meanwhile, technically savvy and internationally educated entrepreneurs, with compelling technologies, have easily attracted venture capital for startups that bring niche AI applications to a broader market. Computer vision (CV), for example, is the science for computers or robots to duplicate the way humans perceive and visually sense the world around them. Facial recognition technology, in particular, has been widely used in public security applications, with public security authorities using the technology to spot suspected criminals and even jaywalkers.

In the computer vision sector, Megvii, SenseTime, Yitu Technology, and CloudWalk are collectively referred to as China's "Four CV Dragons". The Four Dragons' growth coincides with China's full embrace of facial recognition technology and its integration into the daily lives of the Chinese population. In recent years, CV companies have also used the AI technologies to transform a variety of industries from finance to entertainment, transport to healthcare, and more.

Megvii focuses on face detection, recognition, and analysis across different platforms. Its core product Face++ is a cloud-based face recognition technology platform, and it has been widely applied in various industries. Through a partnership with Alibaba, Face++ has been integrated into Alipay (the widely used mobile payment platform) to support facial scan logging in and Smile to Pay, a payment method that allows users to make a purchase by scanning their faces. Face++ is also integrated into smart city applications, where it's deployed to optimize traffic flows and "see" incidents that require police or medical attention. In line with the AI product's name, the Megvii technology is similarly used by Meitu, the popular beauty-enhancing photo editing application. (The Meitu, which means "beautify pictures" in Chinese, app offers features that can remove wrinkles, smooth pores, and lengthen legs in photos.)

SenseTime specializes in deep learning-enabled computer vision technologies, such as facial recognition technology that can be applied to payment and picture analysis for bank card verification and security systems. SenseTime has supplied automatic face scanning systems to many railway stations and airports across China, with a near perfect accuracy rate. For example, it has signed agreements with China's largest subway operator, Shanghai Shentong Metro Group, to use AI to monitor metro traffic. At US$7.5 billion (according to media reports in late 2019, following a recent round from investors, including SoftBank Group), SenseTime is the world's highest-valued AI startup. According to *South China Morning Post*'s 2019 internet report, SenseTime and Megvii are major exporters of AI solutions for security and surveillance in government and commercial markets across Southeast Asia, Latin America, and Africa.

Yitu Technology operates a cloud-based visual recognition engine that enables computers to detect and recognize faces and cars. The system has been mostly used in surveillance and crowd-tracking, and the company's clients include state authorities such as China Customs and China Immigration Inspection. **Cloudwalk** has also been the leading AI tech supplier for China's banking industry.

Natural language processing (NLP) technology is another example. **iFlytek**, China's leading speech-recognition company that creates voice recognition software, has changed the way doctors write up medical records, thus giving them more time to spend with patients. Traditionally, doctors have to spend much of their time writing up patients' medical records every day. They usually write them between surgeries and very often stay in the office after working hours to finish the paperwork. iFlytek's speech-recognition application was tested at hospitals, where doctors could record their diagnoses vocally at "considerably high accuracy".

iFlyteck has also developed more than 10 voice-based mobile products covering education, communication, music, and intelligent toys. It has won a series of worldwide speech and AI competitions, including the Blizzard Challenge, reportedly the most authoritative international competition in speech synthesis. As such, iFlyteck was named by China's MOST (Ministry of Science and Technology) to be the national AI champion for speech recognition, along with **Alibaba** Group (smart city initiatives), **Baidu** (autonomous driving), Tencent (computer vision in medical diagnosis), and SenseTime (intelligent vision).

Collectively, the established internet firms and tech start-ups are investing billions in building new research centers, hiring experienced AI experts and young data scientists, and even setting up labs in the US' Silicon Valley to work on the latest algorithms, smart robotics, and self-driving cars. Tencent even created its own version of a Go-playing AI program called *Jueyi* (or FineArt in English), but only after AlphaGo had already retired; hence, there was no chance for the two AI programs to meet in a direct contest. Nevertheless, the AI race between Chinese and US tech companies is on.

5G iABCD New Infrastructure

If Chinese companies' embrace of AI has been resolute and aggressive, the decisive commitment from the Chinese government, which has announced a sweeping vision for AI and digital economy excellence through a series of policies and initiatives (see Table 1.2), is truly extraordinary. Perhaps it was a

Table 1.2   Major Chinese national and local government strategy and policy initiatives relating to AI and the smart internet (2013–2020)

Name	Issuer	Date	Focus
Promote "**Information Consumption**" to Increase Domestic Consumption	State Council	August 2015	To use information technology to promote more domestic consumption (such as e-commerce)
"**Broadband China**" strategy	State Council	August 2015	To significantly improve the country's information infrastructure, including 4G network coverage to cover all the cities
"**Internet+**" Action Plan	State Council	May 2015	To develop internet access for industries, and leverage AI to create new services and applications
"**Internet+**" and **AI** Three-year Action Plan	NRDC	May 2016	To develop a fundamental AI ecosystem and produce world-class AI enterprises
Next Generation **Artificial Intelligence** Development Plan	State Council	July 2017	Setting out roadmap for China to reach global AI leadership position by 2030
Three-year Plan for **Next Generation AI** Development (2018–2020)	MIIT	December 2017	Setting near-term guidelines based on the national AI Strategy
Artificial Intelligence Innovation Action Plan at Higher Education Institutions	MOE	April 2018	To establish at least 50 AI academic and research institutes by 2020
"**Intelligence Plus**" Imitative	State Council	March 2019	To implement AI to transform and upgrade manufacturing industries
Guidelines for Market-based Allocation of **Production Factors**	Communist Party of China Central Committee and the State Council	March 2020	Defining data as a new type of production factor, on par with land, labor, capital and technology
New Infrastructure Initiative	State Council	May 2020	Calling for quickening the pace of new infrastructure investment for digital economy

Note: **State Council** – China's Central Government; **NDIC**—National Development and Reform Commission; **MIIT**—Ministry of Industry and Information Technology; **MOE**—Ministry of Education.

coincidence of timing, but soon after the Wuzhen Go match, China's central government released "A Next Generation Artificial Intelligence Development Plan" in July 2017.

The plan calls for: (a) a homegrown AI to match that of Western developed countries within three years (by 2010), (b) China's researchers to make "major breakthroughs" in AI theory by 2025, and (c) Chinese AI to be the world's undisputed leader ("occupy the commanding heights") by 2030. This plan probably is the first in the world and by far the most ambitious move to react to the AI revolution.

To solve the shortage of AI talent, China's Ministry of Education in April 2018 released the "Artificial Intelligence Innovation Action Plan at Higher Education Institutions", which envisions that by 2020 China would establish at least 50 AI academic and research institutes for 100 interdisciplinary majors that combine AI with traditional subjects, such as mathematics, statistics, physics, biology, psychology, and sociology, among other disciplines. Governments across the world are rushing to support innovation in AI, but none has published as concrete a plan as China and—more importantly—to execute on.

It should be highlighted that China's digital economy push started much earlier than the ambitious AI plan. It started in August 2013, when China's State Council issued a blueprint to officially promote "**Information Consumption**". At the beginning, the term was mostly related to "consumption-based on information technology", with e-commerce based on internet channels as the focus. Soon the concept was expanded into "quality information products for consumption", such as movies and online videos, which led to a boom of online entertainment. In 2015, the "internet plus (+)" strategy was unveiled in the annual government work report, encouraging traditional industries to use the internet to run key aspects of their business and find new business models.

As a result, China's mobile infrastructure has developed with remarkable speed (see Figure 1.3). According to China internet Network Information Center (CNNIC)'s most recent

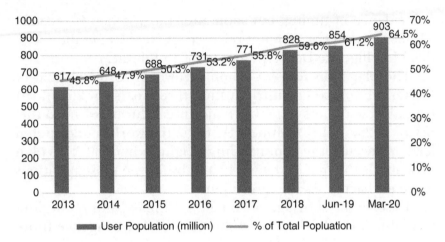

Figure 1.3   China's internet population jump with mobile internet
Data Source: CNNIC, March 2020.

annual report, March 2020, China had the world's largest internet user population of 903 million, representing a penetration rate of 64.5% (of a 1.4 billion population) and a 50% increase from the internet user number of 2013. The percentage of those using mobile phones to go online exceeded 99% (897 million). Since 2013, China has been the largest e-commerce market in the world. (For the mobile economy's early development, refer to the author's 2016 book *China's Mobile Economy – Opportunities in the Largest and Fastest Information Consumption Boom.*)

In recent years, new policies have a much broader goal of digitalizing the whole economy (instead of only the internet sectors) with the latest digital and data-driven technologies (beyond internet connectivity). In 2017, **"Digital Economy"** as China's future economic model was put into the government work report for the first time (also referred to as "Digital China"), and the AI flavor was soon added in 2019 with the notion of **"Intelligence Plus"**. The market expects far more profound transformation of China's economy to come from "intelligence +" than from earlier years of "internet +", and that's also why the next phase of development is referred to as the "second half game" of the mobile internet economy.

The State Council's 2019 work report highlighted the importance to "create **industrial internet** platforms and expand Intelligence Plus initiatives to facilitate transformation and upgrading in manufacturing". The industrial internet concept involves the broader adoption of advanced consumer and industrial applications that are powered by digital infrastructure and data analytics, which is in line with China's wider ambitions to lift its industries up the value chain and better compete globally in emerging technologies, dubbed the Fourth Industrial Revolution by the World Economic Forum (WEF).

Further, in 2020, China's central government issued an economic policy guideline to include "data" in "factors of production", joining the traditional elements of land, labor, capital, and technology, for which the government will take key measures to accelerate the cultivation of the "**data market**". Under this framework, personal data management and protection, unified industry data standards, public and private data sharing, among other key issues of the digital economy, are seeing accelerating development. (Data law developments are discussed in Chapter 8.)

Eventually, all these concepts are summarized by the "**New Infrastructure**" initiative in the Government Work Report, delivered by Premier Li Keqiang in May 2020, which aims to vastly enhance internet infrastructure to spur digital consumption and encourage the development of apps for online working, distance learning, telemedicine, vehicle networking, and smart cities (see Figure 1.4). "We will step up the construction of new types of infrastructure. We will develop

Figure 1.4   From "information consumption" to "digital economy"

next-generation information networks and expand 5G applications. We will build more charging facilities and promote the wider use of new energy automobiles. We will stimulate new consumer demand and promote industrial upgrading," said Premier Li.

This new infrastructure push comes from the central government, local provinces, and related ministries like MIIT (Ministry of Industry and Information Technology). In total, China is expecting, over the next five years, to top more than 27 trillion RMB (close to US\$4 trillion) in new infrastructure construction and related investments, according to Haitong Securities. Compared to traditional infrastructure, such as roads, railways, and bridges (which were the main form of investment stimulus in China during the previous global financial crisis in 2008–2009), the "new" represents infrastructure built on advanced digital technology—it's infrastructure that's data driven.

At the center of new infrastructure is the next-generation (5G) wireless networks for data transmission. The 5G technology's most visible advantage is its data transfer speed, which is expected to be 1000 times more powerful than 4G technology. The high-speed and highly stable 5G services are expected to encompass wireless applications far beyond basic internet communication to include smart cars and advanced manufacturing. Ultimately, 5G is for the "Internet of Things" (IoT), a loose term used to describe a network of mobile internet, smart devices, home appliances, and any physical objects through cloud technology. For example, the faster data transmission by the 5G technology is expected to provide a constant and reliable stream of real-time street data for driverless cars to function effectively and safely.

Based on the superfast cellular 5G networks, advanced applications of AI, blockchain, cloud computing, and Big Data analytics are expected to make major breakthroughs. Together, they form the tech infrastructure of China's digital economy, which can be summarized as 5G iABCD (see

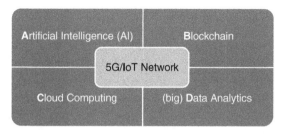

Figure 1.5   New infrastructure – 5G iABCD

Figure 1.5). Companies from every sector of the economy are set to embrace this "full suite" of information technologies for enterprise purposes.

Just like AI and blockchain, China's Big Data and cloud computing market is growing at a rapid clip. In recent years, data centers—secure facilities that house large-capacity servers and data storage systems—have been built at a breathtaking pace to catch up with the explosive demand of data services. The latest from the Chinese government is to build "industrial big data" centers nationwide, such that massive amounts of "industrial big data"—mostly manufacturing and production data—can be utilized to advance the capabilities of traditional industries.

These data facilities are used by enterprises to remotely store large amounts of data, manage their business applications, and host cloud computing operations. Thanks to the cloud units of internet companies like Alibaba and Tencent, China has emerged as a global leader in data-intensive computing. Alibaba Cloud, for example, was formed to provide cloud services to Alibaba's business network, and increasingly external enterprise customers. As the following two sections show, the digital platforms not only help industrial companies' digital transformation, but also provide an important infrastructure for tech startups' innovation. In short, the Chinese internet companies that emerged from the mobile internet boom are becoming the technology infrastructure of China's digital economy.

Digital Transformation in the Cloud

Regarding China's new infrastructure and enterprises' digital transformation, Alibaba Group and its Alibaba Cloud are a perfect example. During the mobile revolution, all Chinese companies started to digitalize their business data and internal systems, and they are increasingly migrating operations and data to the cloud infrastructure (until recently, most of them still rely on local computing in their own data centers). Though Alibaba is mostly known for its e-commerce empire and Singles Day, it has become the first company in the world to run all core business systems on its own public cloud platform. Moreover, its fast-growing cloud computing business further supports other Chinese businesses' cloud-migration journey.

For background, Cloud technology (also known as "cloud computing") means that servers, data storage, databases, networking, software, and analytics are hosted on the internet and stored on large, privately owned data centers. The cloud provider enables end users to "rent" and remotely access IT resources from cloud providers on a pay-per-use basis, which provides an efficient alternative to the local hosting and operation of IT resources. Businesses subscribe to cloud services and pay either a monthly or annual fee, just like buying electricity from a power grid. This fee is determined by the amount of data and number of users a business requires – making it easier for a company to scale its operations up and down. Overall, cloud computing provides both improved system scalability and cost savings over traditional IT infrastructure.

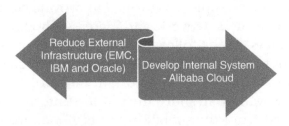

Figure 1.6   Alibaba's IT infrastructure goes internal

As Alibaba's e-commerce businesses have enjoyed explosive growth in recent years, an upgrade of its own internal systems was much needed. Rather than continuing to procure expensive equipment and services from **Oracle**, **IBM**, and **EMC**, Alibaba began replacing its existing IT infrastructure with in-house systems and migration to the cloud. Consequently, Alibaba Cloud was formed to provide cloud services to Alibaba's business network. Today, Alibaba Cloud is also Alibaba Group's cloud computing business for external corporate customers, offering a complete suite of services such as database, storage, network virtualization services, Big Data analytics, elastic computing, large scale computing, machine learning, and IoT services.

Alibaba Cloud has played a major role in educating the domestic market about the benefits of cloud computing, and it maintains its leadership position by developing solutions that enable the digital transformation of businesses across industries. During the last quarter of 2019, it reached two important milestones. First, the cloud computing business for the first time generated over 10 billion RMB of revenue (about US$1.5 billion) in a single quarter. Second, ahead of the 11.11 global Shopping Festival in 2019 (see the details in Chapter 2), Alibaba Cloud enabled the migration of the core systems of Alibaba e-commerce businesses onto the public cloud. This internal migration is a major milestone that not only generates greater operating efficiencies for Alibaba, but also encourages more companies to adopt its public cloud infrastructure.

According to Alibaba, the performance, reliability, and agility of the core ecommerce platform increased drastically after completely moving onto its public cloud. ("Alibaba is like a high-speed jet that is capable of upgrading its engine during the course of flight", one Alibaba executive said.) The next step for the company is to share Alibaba experience with its customers. As such, the investment and adoption of the cloud in China are growing rapidly. According to Canalys' statistics in March 2020, Alibaba Group is China's largest provider of public cloud services, taking almost half of the market.

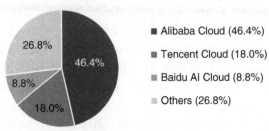

Figure 1.7   Alibaba Cloud leads the Chinese market
Data Source: Canalys, March 2020.

(See Figure 1.7. Tencent and Baidu, two major internet companies, are also sizable cloud service providers.) According to Gartner's April 2020 report, Alibaba Group is the world's third largest and Asia Pacific's largest Infrastructure as a service provider by revenue in 2019.

As the Alibaba Cloud case illustrates, Chinese internet companies have turned themselves into tech companies, and they subsequently help companies from all sectors of the economy to digitally transform. Across just about every industry sector, Chinese companies are investing heavily in research and development of the latest digital technology. The consensus is that digital revolution is more than simply advancing e-commerce through mobile platforms. The more profound value creation is to find incremental consumption demand and new business models, where traditional industries use new set of digital tools to run key aspects of their business.

In other words, digital transformation is more about "digital thinking" than "internet adding". The companies will become "smarter" by fusing institutional human knowledge with machine learning, and their increased efficiencies, faster decision making, and cost savings will all lead to better customer experiences.

For example, the traditional retail industry has been completely remodeled, thanks to the largest mobile e-retailing market in the world. Retailers are forced to move away from the out-of-date perception that e-commerce is merely another sales channel for their products, as it is critical for customers to get the same products, services, and shopping experience in every channel where they choose to make purchases. To meet the demand of tech-savvy customers, retailers must provide a "seamless omni-channel shopping experience" by integrating their online and offline channels, instead of having separate systems to sell products online and offline. Hence, the trend of "new retail", where the boundaries between online and offline are disappearing.

During the 2020 coronavirus pandemic, new digital tools have proven to be a lifeline for China's small retailers as the pandemic disrupts business models and consumer behavior. Small merchants—including many with little or no previous online presence—have flocked to live-streaming to boost sales at a time when consumer habits are changing faster than ever. The video streaming has the advantage of enabling live interaction—consumers can send questions and comments that appear on the screen in real time, and sellers can explain and showcase their products in a personal way. In this context, new retail is more about generating demand by exciting people and creating an experience. (The detailed "new retail" cases are discussed in related chapters in Part Two.)

For instance, Chinese shoe retailer **Red Dragonfly** – which closed almost all of its 4000 physical stores due to the coronavirus pandemic—has leveraged video streaming platforms to move its operations online. The store clerks are trained to run regular live streams to reach, virtually, shoppers still wary of confined spaces and to answer customers' queries online much as they would deal with in-store visitors. Foreign brands too have joined in the scramble to engage homebound Chinese consumers. Luxury product maker **Louis Vuitton** hosted a live-streaming sale in March 2020—the first time since the brand entered the Chinese market 30 years ago. Global automakers such as **BMW** have hired professional livestream hosts to introduce their various car models.

For every business sector, the transition from an industrial economy that favored mass production and scale (which is "product driven") to a digital economy that favors user connectivity, data, and their personal preferences (which is "user driven") is challenging the very nature of the "core competence" long held by the sector and its leading companies. What's promising is that the tremendous revolution in e-commerce and retailing could be similarly applied to other industries when they embrace the digital transformation:

- **Responsive to market demands**. New applications are driven by the needs of the market, aiming to solve the "pain points" of existing services and practices.
- **Keeping agile for new product development**. Companies adapt their offerings quickly based on the "local needs" of Chinese consumers.
- **Deep integration of online to offline businesses**. Online experience is not enough (a shocking concept after many years of mobile e-commerce boom) and must be complemented with offline services.
- **Emphasize a platform instead of single product offering or a specific vertical business**. An ecosystem offering multiple connected products and services is more resilient.

For example, the Chinese insurance market is witnessing a remarkable pace of innovation, which started with the basic "internet +" before undergoing a shakeup of the industry. The insurance companies have moved through four phases in their digital evolution (see Figure 1.8), resulting in more direct engagement with customers and more tailored products for

Figure 1.8   From "Internet +" to data-driven digital economy

them. Furthermore, smart technologies are now revolutionizing the core operations of insurers, such as damage assessment for car accidents. The four phases are as follows:

The first phase involves setting up online channels to reach both customers and insurance agents and automating back-end processes, including underwriting and claim processing. **The second phase** involves creating digital services to help customers with related offline activities, such as making a doctor's appointment through a mobile app. Through the first two steps, companies increase the number of customer interactions and strengthen their connection to customers. **The third phase** then involves consolidating customers' online and offline data to provide more personalized solutions based on the improved understanding of customers.

The rewards are significant for issuers with successful digital implementations. Not only can they engage directly with customers to provide seamless interactions on products and services, but also gain access to comprehensive customer profiling information to win in long-term competitions on customization. According to Tencent's statistics in 2018, the online insurance purchase in China had an 18-fold increase in the past five years, covering a total number of 220 million insurers.

The fourth and latest phase is the most exciting, which involves using smart technology and data analytics to revolutionize the core insurance business. For example, AI has been successfully put into vehicle insurance. Alibaba's financial arm, Ant Group, has developed Ding Sun Bao (Damage Assessment Tool), which aims to standardize damage assessment and make it more objective, reducing the potential for human adjusters on the scene to be influenced by the drivers involved. This digital tool is already adopted by large insurers, including China Taiping, China Continent Insurance, Sunshine Insurance Group, and AXA Tianping.

Ding Sun Bao can remotely collect photos of external vehicle damage, analyze them using AI-driven, deep-learning image recognition technology, reconstruct the scene of the accident, and assess the damage. One advantage is it makes damage assessment more objective, reducing the risk that claim handlers on the scene could be influenced by drivers involved in an accident; hence, there is a higher degree of accuracy in damage assessment. Another advantage is that through machine learning, it can process claims within a few seconds, much faster than human adjusters alone could handle.

At the first launch of DingSunBao in June 2017, Ant Group set up a challenge between "six experienced human claims adjusters" and its AI program. Each team evaluated 12 cases, during which DingSunBao took six seconds to assess the damage and set claim amounts, whereas human adjusters took 6 minutes and 48 seconds to reach their conclusions. Both judged that one of the 12 cases required further investigation. In May 2018, Ant Group launched version 2.0 of its DingSunBao app, which was upgraded into a video-based AI application.

With the upgraded version, car owners can use their smartphones to capture video clips of their cars for submission. Vehicle damage information is displayed automatically, including where and how to repair the vehicle and how much the car owner can claim from insurers, saving time in filing claims and offering transparency on what's likely to be covered. According to Alibaba, this new version's secret sauce includes 46 patented technologies, such as simultaneous localization and mapping, a mobile deep-learning model, damage detection with video streaming, a result display with augmented reality, and others. More industry transformation is coming as Ant Group is now using blockchain technology as the new infrastructure of insurance claims processing.

In 2020, the COVID-19 outbreak gave a digital transformation lesson to almost every industry. From a technology perspective, the outbreak has given a push to many new technologies such as drones and robots, as authorities seek to better detect and respond to the disease. Autonomous robots have replaced human cleaners at segregated wards. In rural areas, local authorities are using drones equipped with loudspeakers to alert citizens seen in open areas without face masks. In particular, it accelerates Chinese companies' shift to the cloud as enterprises try out cloud initiatives to facilitate remote working and remote interactions. For companies managing their internet infrastructures, making adjustments to computing needs on the fly is expensive and complicated. Cloud computing makes it easier.

On the other hand, many pandemic-driven new business practices will remain part of daily life. For example, live-streaming is now an important shopping channel, with or without the "home stay" order, and the digital platforms are developing enhanced services for livestream hosts with new digital technologies, such as 5G and AR- (Augmented Reality-) powered features. With a strong sense of urgency, all companies are rushing to learn how new 5G iABCD digital technologies can be integrated into their businesses to unlock value from nontraditional angles. The digital transformation of China is accelerating.

BaaS Startup Innovation Ecosystem

With government endorsement in the background, a dynamic ecosystem of entrepreneurs and startups is organically being built up and rapidly expanding (see Figure 1.9). The network of established internet firms and their seasoned entrepreneurs, endless eager talents, abundant angel investors and venture capital, and a sophisticated manufacturing system are collectively making China one of the most interesting centers of innovations in the world.

The core of this ecosystem is a network of "graduated Chinese entrepreneurs" from established tech firms at home and abroad. Around their passion for new startups, there seems to be an endless supply of venture capital, young talent, and

Figure 1.9   China's innovation ecosystem

government support. This development resembles the multiplying effect seen in the Silicon Valley ecosystem in the last few decades, where generations of innovators, from Intel, Netscape, Google, and PayPal, have created waves of startups. Whereas the previous generation of China internet companies was mostly copycat versions of Western sites, the new generation startups are creating a remarkable wave of innovation that challenges the long-held perception of "Made in China" copycatting.

Among 5G iABCD (see Figure 1.5), the blockchain is the latest addition to the startup scene. Thanks to fresh endorsement from China's central government, the blockchain industry is experiencing a boom. According to the China Internet Report 2020 by SCMP (*South China Morning Post*) Research, about 80% of the top tech companies recognized by the SCMP have either run a blockchain project or invested in blockchain technology. Scores of blockchain startups are popping up daily, even the coronavirus hasn't been able to keep them down.

For example, because food scandals—ranging from rice grown in cadmium-tainted soil to infant formula tainted with melamine—is widespread in China, blockchain is used to support supply chain transparency and auditability to reassure consumers wary of food safety. In June 2019, Walmart China entered into a partnership with consulting firm PricewaterhouseCoopers (PwC) and blockchain firm **VeChain** to create the Walmart China Blockchain Traceability Platform.

VeChain is a Shanghai-based blockchain provider specializing in governance and business ecosystems. They have created a tokenized public chain called VeChainThor. The companies work together to provide real-time traceability throughout the supply chain, which has traditionally been challenging due to fragmented data sharing systems that are often paper-based and can be error-prone. Built on the VeChainThor blockchain, the traceability platform enables shoppers to track consumer goods back to their source, inspection, and shipping.

Amid the coronavirus pandemic, startups like Hyperchain launched a blockchain-based platform to improve donation efficiency. In February 2020, the blockchain startup teamed with China **Xiong'an Group Digital City Corporation** (as the name suggests, it is the company that operates the blockchain city of Xiong'an) to create a donation tracking platform called Shanzong, which means "trace of good deeds" in English. At a time when some traditional charities like the Red Cross have come under fire for poor distribution of resources to people in need, Shanzong tracks what kind of donations are given, from money and masks to medical materials, how they have been matched to areas of need, and when they have been delivered, according to a Hyperchain statement.

In fact, China is shown as dominating in the 2019 Global Blockchain Invention Patent Ranking report from IPRDaily, a Chinese intellectual property news site. It would appear that China is predominantly driving the world in the utilization and development of blockchain technology. According to the report, China accounted for the top three spots, seven of the top 10 and 19 out of 30, and the Chinese companies

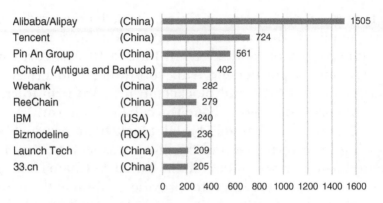

Figure 1.10   Global blockchain patent application 2019 ranking

have represented about 70% of the world's global blockchain patent applications (see Figure 1.10). Alibaba and its financial arm, the Ant Group, have by a wide margin filed the most patents identified with blockchain on the planet.

With an initial public offering (IPO) announced in August 2020 and targeting a valuation of more than $200bn, the Ant Group is the world's highest valued financial technology (or fintech) company—and most valuable private tech startup globally. Its Alipay is China's largest mobile-payment network, spanning 900 million users, and it also operates one of the country's largest investment products platforms, lends to individuals and small businesses, and operates a credit-scoring system based on customer data analytics. (See the detailed case study in Chapter 5. The company was known as Ant Financial until June 2020, when its official Chinese name was changed to "**Ant Technology**." A spokesperson said the company wants to be referred to in English as "**Ant Group** Co.", better reflecting its role as "an innovative global technology provider" to businesses.)

Of course, most blockchain protocols are open source. Therefore, the ranking of blockchain patent applications is not directly equivalent to blockchain technology innovation power. Nevertheless, the increase in the number of blockchain patents demonstrates a developing interest by China and business organizations like Alibaba, which has been on the top since 2017. Furthermore, Alibaba, Tencent (ranked No. 2), and other firms also provide blockchain platform infrastructures for the market (BaaS – blockchain as a service). Their BaaS services provide an important link between blockchain architecture and the enterprise-level blockchain projects.

These BaaS services enable blockchain entrepreneurs to focus on building new applications, leading to a wave of young blockchain companies in China. For example, Tencent is collaborating with Waterdrop, a crowdfunded health insurance firm, to develop a medical and insurance solution leveraging blockchain technology. Tencent plans to integrate the solution into its WeChat messenger, which will help more than one billion Chinese users to access their medical bills conveniently and securely. The solution will also benefit medical institutions and insurance firms by facilitating an efficient billing system to prevent claims from fraudulent invoices.

In the coming years, the pace of innovation will continue accelerating. For one thing, established tech firms, as well as up-and-coming companies, serve as useful infrastructure for next generation tech startups. For another, the synergies of 5G iABCD are gradually taking shape. The different technologies can feed into each other and create an ecosystem of automation—IoT devices collect data on millions of criteria, which is then collated in the cloud and managed by Blockchain, analyzed by Big Data, and used to train and improve AI algorithms for real life applications. As 5G iABCD technologies interact and improve each other, the huge synergies will spur more innovation in China (see Figure 1.5).

Splinternet and the Digital Silk Road

China's rapid digital transformation has had profound implications for global stakeholders dealing with the China market. Foreign investors are richly awarded, and consumer goods companies see an emerging market filled with opportunities from an expanding middle class. Overseas users are adopting smartphones and mobile apps created in China, but Silicon Valley tech giants are taking notice of new competition arising from Asia. Today, many Chinese companies are looking to expand overseas, and their impact is increasingly felt by foreign investors, consumers, startups, and industry companies across the globe.

Their overseas expansion has accelerated since China started the "Digital Silk Road" (**DSR**) initiative in 2017. The DSR is the new digital dimension of the "Belt and Road Initiative" (**BRI**), which was launched by the Chinese government in 2013 to promote cross-continent trade through major infrastructure investments. Whereas BRI has been commonly associated with physical infrastructure projects, such as roads, railways, energy pipelines, and ports, the DSR will bring advanced IT infrastructure to the BRI country, such as broadband networks, e-commerce hubs, and smart cities.

In DSR, China sees its burgeoning digital economy as a success story that can be shared with emerging markets, and it has also become a partner for other countries' digital revolution (see the detailed DSR discussion in Chapter 10), including examples such as:

- **Localized solutions**. Baidu has opened a language processing lab in Singapore to improve its search applications for speakers of Southeast Asian languages, and it has launched a localized version of its search engine in Brazil.
- **Smart infrastructure**. Alibaba Cloud has set up data centers in Indonesia, offering a reliable and cost-effective cloud product and services to Indonesian businesses, particularly small to medium-sized enterprises (SMEs) and startups;
- **Education.** Alibaba set up worker training programs in Malaysia to teach small and medium-sized enterprises to sell their products on Alibaba platforms;

- **Investing into new startups**. Tencent and Didi Chuxing, among others, have invested in the ASEAN car sharing apps, including Grab and Go-Jek, which surpassed the global ride-hailing service Uber in Southeast Asia. (Notably, Chinese capital has also actively invested in Silicon Valley, but in recent years, such investments have plummeted due to tightened US CFUIS investment regulations.)
- **Global M&A transactions**. In early 2018, a Chinese investment consortium acquired a majority stake in London-based data center operator Global Switch, in a series of transactions worth more than US$3 billion. Just like investing into new startups, these M&A deals aim to find synergies between overseas companies and the Chinese market.

Not surprisingly, the latest addition to the DSR is, again, the blockchain. In April 2020, China launched the Blockchain Services Network (**BSN**), which is a critical part of China's national blockchain strategy that was announced by President Xi in late November 2019. The BSN is an ambitious effort to include as many blockchain frameworks as possible and make them accessible under one uniform standard on the BSN platform. As such, it's the largest blockchain ecosystem in China, and in the DSR context, it is rapidly expanding its network overseas.

According to its official announcement, the BSN promotes low-cost development, deployment, and maintenance of consortium blockchain applications. Just like building a simple website on the internet, developers can deploy and operate blockchain applications conveniently and at extremely low cost. So far, no blockchain project has yet found widespread commercial usage globally. By significantly reducing entry barriers for blockchain application developers, the BSN can potentially drive significant innovation in traditional businesses across countries, which would have profound implications for cross-border applications from finance and payments to commerce.

As China's digital economy gains increasing global influence, the global economy sees China in Asia and the United States in the West forming two leading innovation centers of the world. Chinese and US companies represented 90% of the total market value of the top 70 digital business platforms, according to the United Nation's 2019 digital economy report. The two innovation centers have different strengths, but are also highly connected—the United States is the incubator of original technologies, and China is the best market for commercialization. Thanks to its largest user population, China excels at incremental innovation, but still lags the United States in transformational, science-based innovation.

Until recently, the two markets enjoyed a cross-border flow of ideas, capital, and talents. Silicon Valley operated on three key ingredients: code, connections, and cash, and the Chinese link brought in abundant supply (before the recent US–China tech and trade tensions). Chinese engineers churned out plenty of the first, and Chinese venture capital brought an infusion of the second and third (see Figure 1.11). (As a cross-border investor, the author used to focus on US investments that had a strong potential synergy with the Chinese market from a commercial perspective as well as Chinese portfolio companies that sought to globalize in overseas markets.) The two-way bridge

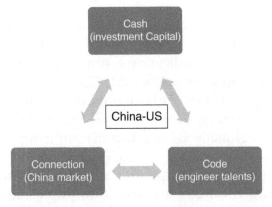

Figure 1.11   The 3C transpacific synergy (of yesterday)

created synergies and speeded up startup launches, innovation, and scale on both sides of the Pacific Ocean.

However, investment from China into US tech companies quickly slowed in August 2018 after the Trump administration stepped up vetting of deals over national security issues in American critical technologies. One key driver is that, with focused policy and substantial investment by its government, and the unrivalled user data pool, original 5G iABCD innovation is advancing in China, which, in more and more fields, is in head-on competition with the United States. Another important factor is, for each country, the data used to train AI models are locally managed and carefully guarded, as tech innovation is now of national security and geopolitical significance.

Even after the US–China phase I trade agreement ("a truce") was reached in early 2020, the tech war has shown no sign of easing. For example, stoked by the COVID-19 pandemic, the short video app TikTok was second in downloads to the Zoom video-conference app in the first half of 2020, according to market-research firm Sensor Tower. Since 2019, US lawmakers have been calling for an investigation of TikTok's relationship with its Beijing-based parent company (ByteDance) and the Chinese government. In July 2020, Secretary of State Mike Pompeo suggested that the United States is considering banning TikTok because it views the popular social media app as a security threat – despite the fact that TikTok had spent much of the previous 12 months trying to distance itself from its Chinese roots.

And the blow came more quickly—and broader—than everyone anticipated. Weeks later, on August 6, US President Trump issued sweeping bans against TikTok and fellow Chinese tech app WeChat (the "superapp" owned by Tencent), citing concerns that TikTok and WeChat collect "vast swaths of information from its users"; i.e., from US users. On top of that, Trump issued another executive order within a week, ordering ByteDance to divest the US operations of TikTok within 90 days.

In response, ByteDance and TikTok filed a lawsuit against the US government, challenging the president's executive order that bans US companies from doing business with them on national security grounds. According to a ByteDance statement, that executive order was issued "without any due process". Noteworthy, TikTok's lawsuit did not challenge the order to divest its US assets, and Microsoft and Oracle and other private investors immediately started a bid for TikTok's businesses in the US, Canada, Australia, and New Zealand, in the hope that the TikTok platform could bring significant synergies to their existing consumer-facing businesses.

Regarding the popular messaging app WeChat, the order would bar "any transaction that is related to WeChat". In response, a group of WeChat users formed the US WeChat Alliance Group, a nonprofit group (not affiliated with the app's owner Tencent), which then filed a lawsuit against the Trump administration's executive order, calling the prohibition unconstitutional. Interestingly, because the WeChat app is ubiquitous in China, more than a dozen major US multinational companies raised concerns in a call with White House officials that such a ban could undermine US companies' competitiveness in China, according to *Wall Street Journal* reports.

Compared to mobile apps popular among teens, the advanced tech industry is the big boy version of that. Huawei and ZTE, China's two major telecommunications systems makers, have been banned from buying US microchips for their 5G network push. More recent additions to the trade blacklist are eight Chinese artificial intelligence companies, including AI national champions SenseTime, Megvii, and iFlyTek. (As its New Year's message, Megvii said in the first post on the company's official WeChat account in 2020 that being added to the US "Entity List"—which bars it from buying US-origin technology—had actually turned into a "coming of age gift" that had taught the company how to face a complex and changing international environment.)

As a result, these US actions have firmed up China's resolve to cut reliance on US tech smarts and to grow its own core technologies. Relating to microchips, for example, in October 2019, China set up a new national semiconductor fund (its second in less than five years) with 204 billion yuan (US$28.9 billion) – its predecessor was capitalized with US$20 billion in 2014. The new fund has an ambitious goal to cultivate China's complete semiconductor supply chain, from chip design to manufacturing and from processors to storage chips.

What's next for the United States? It's setting up its own sovereign funds to develop 5G network technology, a field in which Chinese company Huawei is the global leader. The battle for digital tech supremacy is on, and this process of the United States and China "designing out" each other's technologies will continue. (See more details in Chapter 9. Also, refer to the author's recent book *The Hunt for Unicorns: How Sovereign Funds Are Reshaping Investment in the Digital Economy*.)

Emblematic of this "decoupling", nothing so far has been as literal and dramatic as a severed data cable that potentially links the two sides of the Pacific. The Pacific Light Cable Network (PLCN), a high-capacity fiber-optic cable project started before 2018, was a joint venture among Google, Facebook, and a Chinese telecommunications company called Dr. Peng Telecom & Media Group, which is the fourth-largest telecom company in China (see Figure 1.12). The cable would run about 8000 miles (13 000 kilometers) and was intended to be a high-speed trans-Pacific data route between the United States

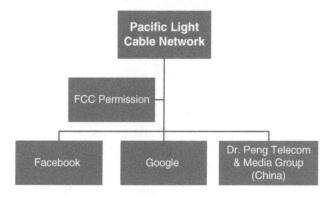

Figure 1.12   The Pacific Light Cable Network

and Hong Kong/China. In 2020, the US government formally denied the Hong Kong link, and Google and Facebook (with Dr. Peng out of the picture) accepted the reality of having to operate just the two fiber pairs owned by the American companies: Google's link to Taiwan and Facebook's to the Philippines.

This is a vivid example of the "splinternet," or separation of Eastern and Western technology worlds, that is coming, and quickly. In the past decade, companies like Google and Facebook have made significant investments into similar cables to handle ever-growing network traffic between the United States and Asia. Now, the national security considerations are changing the way that Internet connectivity across the Pacific is structured. Because subsea cables form the backbone of the internet, by carrying 99% of the world's data traffic, the rules of internet connectivity between the United States, China, and the whole world is being dangerously rewritten.

All in all, a US–China tech decoupling is real and accelerating. Hence, the digital economy is in a vital conflict and crisis: the global tech world, together with at least part of the world economy, is now fractured into two—and potentially more, considering Europe, Japan, and other regions—spheres of influence, whereas tech entrepreneurs are driving the prospect of a technological singularity, hyper-connected society, and internet of everything. The tech companies are only now waking up to the fact that their platforms in the future are going to be a lot less globalized.

Therefore, it is critical for the United States and China, the two tech superpowers, to reach a new equilibrium to collectively lead the future 5G iABCD innovation with the rest of the world. Otherwise, the cross-border flow of capital, talents, and data, which the global economy has taken for granted, is at risk. That is why the Digital Silk Road—at least its concept of global connectivity—is such an important concept to start a global digital economy dialogue. As the G20 leadership together declared at the G20 2016 Summit in Hangzhou, China: collectively, the digital economy revolution will build an innovative, invigorated, interconnected, and inclusive world economy.

2

The World's Largest Mobile Economy

- Ten Years of Xiaomi Smartphones and AIoT
- "Mobile First" and "Mobile Only"
- Shoppertainment: The Digital Middle Class
- Singles' Day: Hard Tech Drives the e-Shopping Festival
- BAT, TMD, and the New ATM

Ten Years of Xiaomi Smartphones and AIoT

The year 2020 marked the 10-year anniversary of Xiaomi, the Chinese smartphone brand. Xiaomi, whose name means "millet" or literally "little rice" in Chinese, became the leading smartphone brand in China in less than five years (2014). The company was valued at US$45 billion after its 2014 financing round, making it the highest valued startup in the world at that time, even ahead of the US$41 billion achieved by the US car-sharing service Uber Technologies Inc. earlier in the same month. In just a decade, Xiaomi has risen from birth to become the fourth biggest smartphone manufacturer in the world.

Since the beginning of the twenty-first century, a large proportion of the Chinese population has stopped using landline phones and moved to using mobile phones. The fact that usage of landlines was not as pervasive as in Western econo-mies meant that, for many, the transition did not involve fixed-line phones at all, but was directly to mobile phone. For many people in China, especially in rural areas, their first internet experience is often mobile instead of a personal computer (PC) – the moment they start using a smartphone.

In 2010, with about US$40 million in initial financing, co-founder Lei Jun teamed with a former Microsoft and Google engineer, Bin Lin, and five other engineers to set up Xiaomi in a small office on the outskirts of Beijing. The company was originally a lean startup looking to sell phones at cheap prices over the internet. The company also started a software plat-form named MIUI for the phones, which were adapted from Google's Android system. In August 2011, Xiaomi introduced its first smartphone, Mi-1, which put Xiaomi decidedly late in the hardware competition (for a reference point, Apple was releasing the iPhone 4S by then).

Focusing on a niche ignored by premium brands like Apple and Samsung, Xiaomi's low-priced, high performance devices have played well into the general Chinese population's passion to own one smartphone and to access the internet for the first time. Xiaomi has been generous in offering top-notch metal material, screen resolution, chip processor, camera, and other features, so its low price is associated with a quality brand instead of a "cheap" product. From time to time, Xiaomi was referred to as "the Apple of China" for the excitement it generated among Chinese consumers. Its similar design to the iPhone almost makes it an alternative to the Apple brand for many Chinese consumers.

Examining the overall design of Xiaomi's phones reveals a shape and style with some degree of similarity to Apple's iPhone. From time to time, the company has been criticized for some of its design borrowings from Apple and other rivals (such as Samsung). Some critics suggest that XiaomI has taken advantage of Apple's high-end brand identity in China: many Chinese consumers have idolized the Apple brand, and Xiaomi provides a similar product at a much less expensive

Table 2.1   2019 Global top 5 smartphone companies

Company	Market Share (2019)	Market Share (2018)	Year-on-year Increase (decrease)
Samsung	21.6%	20.8%	**1.2%**
Huawei	17.6%	14.7%	**16.8%**
Apple	13.9%	14.9%	**−8.5%**
Xiaomi	9.2%	8.7%	**5.5%**
OPPO	8.3%	8.1%	**0.9%**

Data Source: IDC.

price so that customers are happy to buy a product that could be viewed as an "affordable iPhone".

Traces of imitation of Apple can also be found at the Xiaomi's management team, which does not hide its emulation of Apple and its late founder Steve Jobs. When Xiaomi releases new products, Founder Lei Jun typically comes to the stage in a black T-shirt and converse shoes, not much different from Mr. Jobs's signature outfit. At some Xiomi events, co-founder Lei Jun has used the line "One more thing..." at the end of the presentation, which was the line that Jobs famously used for surprise announcements of innovations at Apple's product introduction events.

Ten years later, Xiaomi has become a serious competitor to Apple. In the list of global top 5 smartphone companies, compiled by the IDC (see Table 2.1), Samsung ranked first in 2019, and Apple and Xiaomi took the No. 3 and No. 4 positions, respectively, with Apple's global market share decreasing on a year-on-year basis, and Xiaomi's increasing. (It's worth noting that two other Chinese brands, Huawei and Oppo, took the remaining spots in the top 5. Together with Xiaomi, these Chinese phones have helped the world's largest mobile internet user population – approximately 900 million – to quickly emerge in China, and they are competing with Samsung and Apple globally.)

But Xiaomi never positioned itself as a smartphone manufacturer. Instead, from the very beginning, the company

insisted that it was an internet company; or, if one had to associate the company with smartphones, a smartphone company
with internet DNA. For the next 10 years, Xiaomi has its eyes
on "upgrade, expand, and global":

First, upgrade. In the market of budget-friendly smartphones (the sub-US$300 mid-range segment), Xiaomi is in
cut-throat competition with more domestic brands. Because
Xiaomi's brand identity was formerly linked with high quality and low price, the risk it faces is that some Xiaomi users
may use Xiaomi phones as a transition product before they
can afford a luxury brand smartphone. The company needs
to either seek high sales volumes at low profit margins (like
the PC market that Lenovo has dominated) or find a way to
transform itself into a premium brand with distinctive features
(as Huawei has successfully done with its Mate series). In 2020,
with the Mi 10 Pro, Xiaomi also stepped into the flagship market with powerful hardware and premium prices.

Second, expand. At the start of 2019, Xiaomi first proposed
its "All in AIoT" initiative, allocating an investment of 10 billion RMB (about US$1.5 billion) in AIoT over the next five
years. (The tech industry is as good at creating sensational
acronyms as Wall Street: "AIoT" is a combination of "AI" and
"IoT".) Essentially, Xiaomi aims to develop a family of smart
home devices that are seamlessly cross-linked under the Xiaomi
"ecosystem", with the Xiaomi phone as a remote control. The
company has already become a brand for consumer electronics like cameras and chargers as well as home appliances like
air purifiers and desk lamps.

As such, Xiaomi describes its future direction as "Smartphone + AIoT", a dual-engine strategy (see Figure 2.1). Because
the 5G connectivity will provide a faster smartphone network,
Xiaomi in its 2019 annual report states that "5G+AIoT" will be
a revolutionary combination that will fundamentally change
"all Xiaomi's products, platforms, and user scenarios", catalyz-
ing the next stage of "our growth in the next generation of the
internet era". From 2020 Xiaomi will increase its investment

Figure 2.1   Xiaomi's dual engine

to at least 50 billion RMB (about US$1.5 billion) over five years in "5G+AIoT", to ensure absolute dominance in the "new smart living era".

Third, global. Xiaomi is a major player, both inside and outside of China, and it keeps growing. Chinese smartphone brands are moving beyond the saturated domestic market to compete on a global stage to maintain their growth; and for Xiaomi, the overseas market is its focus. In the coming years, Xiaomi will strive to boost its market share in places where the company has already established a strong foothold, including India, Europe, Southeast Asia, and Latin America.

India, in particular, is Xiaomi's key market. In the fourth quarter of 2019, Xiaomi was the largest smartphone brand in India by shipments for the tenth consecutive quarter, capturing a market share of approximately 28.7%, according to IDC (International Data Corporation). It also ranked first in India in terms of smart TV shipments for seven consecutive quarters as of the fourth quarter of 2019. (In July 2020, India–China tensions grew over a disputed border area high in the Himalayan mountains, which resulted in a boycott by Indian small businesses of "Made in China". It remains to be seen whether Xiaomi can sustain its leading position in the Indian market in the long term.)

In summary, the 10 years of Xiaomi is a perfect example of Chinese tech companies that have proven their mettle by catching up with global rivals in the smartphone and fourth generation (4G) technology development process. They are now joining a fiercely competitive race—among themselves and foreign players—to become the leading companies to offer fifth generation (5G) wireless networks and iABCD products to global customers. They are moving aggressively into the future of an ultimate multi-device and hyper-connected world; as such, Xiaomi is also a good example of Chinese companies expanding from an earlier internet focus into broader tech-driven digital economy.

Before the 5G iABCD digital economy discussion in the rest of the book, this chapter provides an overview of China's mobile internet revolution over the last few years and how the mobile economy set up the foundation for the forthcoming industrial revolution. (Refer to the author's earlier book *China's Mobile Economy – Opportunities in the Largest and Fastest Information Consumption Boom.*)

"Mobile First" and "Mobile Only"

No doubt, the prevalence of mobile devices in China during the past few years marks an important inflection point in the history of the Internet in China. The spread of low-cost mobile phones quickly reached all parts of this geographically vast country, with an especially profound transformation in rural areas. On the one hand, hundreds of millions of villagers are being linked to e-retailing websites and becoming online shoppers; on the other hand, Chinese villagers are also quickly embracing e-commerce as retailers as the demand for fresh, safe agricultural products grows rapidly in the cities.

In addition to the unrivalled internet user population size, what also makes the Chinese market unique is the fact that China is the largest "mobile first" and "mobile only" market in the world. For many people in China, especially in rural areas, a consumer's first internet experience is often mobile, instead of PC – the moment he or she starts using a smartphone. Therefore, with regard to consumer application, China's market has evolved in a very different way from the Western world, moving more aggressively into mobile. In other words, the Chinese population has leapfrogged into a mobile-first mobile-only era.

For example, the lack of a developed credit card system in China means that mobile payment is the "first" and "only" non-cash payment experience for many users. The large screens strongly preferred by Chinese customers from the outset, a trend driven by the mobile-first tendency, made them great platforms for various types of transaction. In China, the penetration of mobile payments has already surpassed the United States. As many cases in this book illustrate, the mobile payment system has been a foundation for numerous fintech innovations and new business models.

According to findings published by Alipay, Alibaba's mobile payment app and the leading mobile payment channel in China, mobile payment is more popular—to the surprise of many—in China's underdeveloped western regions than in its coastal cities. As early as 2014, the Tibet autonomous region led the country in mobile payment adoption, which was followed by the remote provinces of Shanxi and Ningxia. In these provinces, a higher percentage of online transactions were paid by consumers with mobile devices (as distinct from personal computer channels), way ahead of the cosmopolitan cities of Beijing and Shanghai (see Figure 2.2).

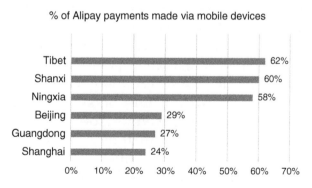

Figure 2.2   Tibet and remote provinces lead mobile payments
Data Source: Alipay, 2014.

In the following years, Tibet remained at the top of the ranking, and in 2016, Tibet was the first province to reach a 90% mobile payment rate. The reason is quite understandable: there is a lack of bricks-and-mortar retail infrastructure and banking system in those regions, so the people there turn to online shopping and mobile payment more frequently for the products they purchase.

Also, the mobile payment infrastructure creates new industries. For example, the internet and mobile devices have changed the way people read novels. Online novels in China are posted by installments, typically a few thousand words every day. People can do the reading whenever they have a few minutes—such as waiting in line or standing in a subway. As a result, a whole new industry has emerged as many online novels are turned into gaming, videos, and movies (as a part of the "Fans Economy", the online literature business is discussed in detail in Chapter 4).

Ride-sharing mobile apps are another product that represent the direction of "mobile only" internet development in China: people can post the information of other shared resources on a PC, but they cannot carry a PC while hailing a taxi on the street. Nor can the driver install a PC in the car. Not surprisingly, ride-hailing apps took off quickly in China, and Uber—the San Francisco-based global car-sharing app leader—saw its China operation expand much faster than it had in the United States. (Uber China was eventually merged into its Chinese rival Didi. See the detailed case study in Chapter 6.)

Because the "mobile first" and "mobile only" trend is transforming every sector of the economy, Chinese companies are more inclined to expand "horizontally" into new sectors, whereas Western companies tend to grow "vertically" to areas upstream or downstream from their original focus. When Chinese companies sense a new opening in the broad industry, they often aggressively create a new business unit to join the competition, or, for the internet and tech giants abundant with capital, "buy" their way into the new businesses through mergers and acquisitions (M&A). In the Chinese market, "horizontal expansion" has led to many innovations with "Chinese characteristics".

For example, when years ago Tencent launched its first instant-messaging product, named QQ, the product was viewed as a replica of the same system on which Yahoo Messenger and MSN Messenger were based. Today, however, WeChat (as well as Tencent's QQ app) have evolved into a much stronger mobile platform than those loosely related foreign counterparts. Mobile commerce users can manage their money, order taxis, and even invest in money market funds, all from their smartphones. By contrast, in developed countries, these activities are not as widespread on mobile or not as integrated on one single mobile app.

Didi Daren ("Thugs for Hire" App)

The imagination of Chinese entrepreneurs seems to know no bounds. In 2015, the widely promoted taxi-hailing mobile app Didi Dache ("Honk Honk Hail a Taxi") unexpectedly inspired a sensational name for a new service app. By changing the character "har" (car) into "ren" (person), the new app was named Didi Daren. Because "da" in Chinese could either mean "hail" or "beat", the app's name could be known as "Honk Honk Beat a Person."

After the app became available for download, some users understandably took the name at its face value. Chinese media reported that people used the app to offer themselves as thugs for hire, providing strong credentials such as "a team of broad capabilities, verified by authorities, and from various professional backgrounds, including sports trainers, retired veterans, experienced thugs, wanted fugitives, and more."

More confusion was created by "Baozou Big News", an online Chinese comedy show that broadcasted a skit of a bullied nurse and a harassed schoolgirl using the app to call for enforcers to beat up the offenders. In the video, the offenders were beaten into unconsciousness, whereas the schoolgirl and nurse thanked the app profusely for "empowering" them. As a result, more people believed that the app was a real service provider for "thug hiring", and the app was downloaded around 200,000 times within three months.

Later, it turned out that the app was a concierge service created by a tech company whose English name was Joke. Instead of being an "Uber for thugs", the app meant to help people to "hail a person" to serve and run errands. The Joke Company played the pun on "Honk Honk Beat a Person" to have a special marketing twist on its product. Amid the fierce competition on the mobile internet, all companies seek every possible edge to grab the attention of potential users. In the case of "Didi Daren", it seemed to prove that it was worthwhile to be audacious.

Before the Didi Daren app was pulled from the app stores due to public controversy, many people actually believed that the app was an innovation on thug-hiring. Some internet pundits made positive commentaries that Didi Daren created a useful platform for average citizens to engage thugs because even people having no relationship with the Godfathers could find the service directly. In addition, at the online platform the price was negotiable, service quality could be reviewed, and word-of-mouth reputation became consolidated. Finally, all the corresponding big data were easy for customers to search.

There were even more sophisticated reviews that linked Didi Daren with new mobile internet-related business models in commercial sectors. They found that the Didi Daren drama illustrated the mobile internet revolutionizing service models by cutting out the "middle-man" or "agent". The conclusion was that the traditional service providers must embrace mobile internet to better serve their customers. When the Joke Company took a free ride on Didi Dache's fame, it most likely had never expected such a profound interpretation of its marketing gimmick.

As a result, the market serves as a giant laboratory. Consumer-facing internet companies can rapidly pilot new products and services to learn what works and what doesn't. Based on market feedback, they could expand, refine, or terminate new offerings quickly. The "shared economy", for example, has benefited from the wild imagination of the China market (see the detailed discussions in Chapter 6). Because of the wide adoption of mobile apps in Chinese's daily lives, in an extreme case, the general public genuinely believed there was a mobile app providing "thugs for hire". (**See the "Didi Daren ['Thugs for Hire' App]" box**.)

Interestingly, "hailing people" has actually materialized amid the COVID-19 epidemic. Chinese companies have come up with a creative way to optimally allocate human resources:

"employee sharing". On the one hand, restaurants are either shut or not entirely reopened, with millions of people staying indoors. On the other hand, the market saw a sharp rise in demand for online businesses, delivering anything from takeouts to groceries and more, along with corresponding demand for more couriers and delivery personnel. Bike-sharing companies have also become an unlikely winner in the pandemic, as people shy away from public transportation to keep their social distance.

As a result, the offline shops of e-commerce platforms, such as Alibaba's **Freshippo** and JD.com's **7Fresh**, and bike-sharing companies, such as **Hellobike**, are temporarily hiring—or "sharing"—thousands of employees from hard-hit businesses such as restaurants, cinemas, Karaoke bars, and film production companies. The workers' original companies continue to pay for their social security and benefits, while the new, temporary employees pay their wages, usually on an hourly basis. The employee-sharing scheme ensures basic income for out-of-work people, and it effectively distributes companies' labor costs.

Since the situation in China gradually nominalizes, the temporary workers start to return to their previous places. But the employee-sharing scheme is to stay for the long

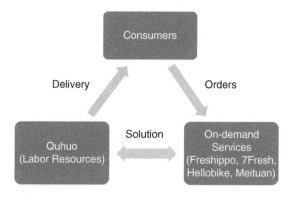

Figure 2.3   Riders for hire

turn, as more and more companies recognize "employee sharing" as an excellent way to put workers where they're most needed. **Quhuo** Limited is the largest "labor sharing" service platform in China, and its corporate name means "fun work" or "fun living", because the Chinese character "huo" could be linked to either work or life. (For the "riders"—the people delivering various services—it may be both.) The platform provides workforce operational solutions for on-demand consumer services, including food delivery, ride-hailing, housekeeping, and bike-sharing (see Figure 2.3). In July 2020, the company went public on the Nasdaq stock exchange.

Shoppertainment: The Digital Middle Class

The mobile internet revolution in China has created an enormous digitally connected middle class. According to McKinsey's estimates, this middle-class group is already the size of the US population and is expected to double within a few years. In addition to its large population size and significant disposable income, this new middle class is also characterized by its quick adoption of mobile applications in its everyday activities. This population uses the internet more for entertainment—text messaging, social network sharing, online game playing, movies and videos streaming, and shopping—than for work, which brings seismic changes to Chinese customer markets.

A list of the most commonly used mobile apps at the end of 2019 provides a perspective on the new generation of consumers' preferences (see Table 2.2). The three most important areas—social network, e-commerce, and video streaming—occupy all of the top 10 spots. As shown in statistics, MAUs (monthly active users) of the top 10 most used mobile apps in China have all surpassed 400 million (see Table 2.2).

Table 2.2   Top 10 most used mobile apps in China

Ranking	App name	Monthly active users (million)	Sector/Company
1	WeChat	970	Messaging app/Social network (Tencent)
2	QQ	763	Messaging app/Social network (Tencent)
3	Alipay	695	Mobile payment (Alibaba)
4	Taobao Mobile	687	E-commerce (Alibaba)
5	iQiyi Video	613	Video streaming (Baidu)
6	Douyin	564	Short video platform
7	Tencent Video	524	Video streaming (Tencent)
8	Youku Video	467	Video streaming (Alibaba)
9	Pinduoduo	443	Social e-commerce
10	Kuaishou	432	Short video platform

Data Source: iiMedia Research, December 2019.

All these apps are discussed in detail in the following chapters. The following is a brief introduction:

First, social network and messaging. Most notably, Tencent's two popular mobile messaging apps **WeChat** and **QQ** take the top two positions in the ranking. Similar to WeChat, QQ is one of Tencent's older applications for person-to-person communication. By combining features from Twitter, Facebook, Instagram, and other social-media services, the newer version of WeChat is now the most popular forum in China. It started as a messaging app, but has evolved into a major mobile commerce ecosystem. Its social network function shapes consumer behavior in significant ways.

In China, social networks have tremendous influence on customers' purchasing decisions. According to A.T Kearney's Connected Customer Study, conducted in the early mobile economy year of 2014, Chinese consumers are 10 times more reliant on social media for purchasing decisions than Americans. This might be explained in part by a cultural difference: social engagement and purchasing behavior are so intertwined in China that Chinese customers tend to seek friends' input before they make shopping decisions. Another cultural aspect could be that Chinese consumers are more skeptical of formal communications, such as quality certificates; instead, they put more value on peer-to-peer recommendations. With almost instantaneous feedback and easy-to-use interfaces, social media platforms are indispensable in the life of Chinese e-consumers.

The eager adoption of social media by Chinese consumers has created unique opportunities for companies that want to gain insights about, and to engage with, the enormous middle class. Therefore, global brands in China, such as Burberry, Coca-Cola, Estée Lauder, McDonald's, Mulberry, Starbucks— just to name a few—have actively engaged the social networks in China (such as WeChat) to develop new advertising strategies. Interestingly, the best example of social network marketing based on WeChat has been none other than the WeChat "red envelope"—the promotion of WeChat's own digital wallet. (See the detailed case study in Chapter 5 relating to fintech topics.)

Second, e-commerce and mobile payments. Alibaba dominates the e-commerce world through two online marketplaces—**Taobao** and **Tmall** (the latter is specifically intended for high-end goods and luxury brands). Its leading position is further strengthened by the widely used online payment system **Alipay**, which is affiliated with Alibaba Group. Mobile e-retailing enables customers to place online orders anytime and anywhere, which aligns with Chinese consumers' desire for speed and convenience of "any time" shopping. **Pinduoduo** is a "group purchase" e-commerce platform, where the users are encouraged to "ask your friends to help cut the price".

Third, video streaming and short videos. Interestingly, half of the top 10 apps are about videos, which reflects the fact that video streaming is accounting for the majority of mobile internet traffic and Is at the heart of online entertainment.

Among them, **Tencent Video**, Alibaba-invested **Youku Tudou,** and Baidu's **iQiyi** are the leading streaming video websites. They provide free professional-made longform video content (PGC) to attract online viewers' eyeballs and rely on advertising income as the main revenue source. **Douyin** and **Kuaishou**, on the other hand, focus on user-generated short videos. Their platforms provide digital tools for the users to create "professional quality user-generated contents" (PUGC). This viral PUGC trend is reshaping the content market with short video marketing strategies becoming increasingly crucial for marketers.

Whereas internet coverage has brought to Chinese customers unprecedented exposure to brands compared to traditional bricks-and-mortar shops, the explosive growth of online entertainment content—with brand images carefully built into the backdrop—presents even more brands for the middle class. For example, in China sparkling wine has traditionally been less commonly used on special occasions or at ceremonies than red wine, but a recent Chinese blockbuster movie series *Tiny Times*—known as the *Sex and the City* of China—provoked new interest in French champagne among the younger generation. The movie series, which depicted the lifestyle of high society girls in their 20s, appealed to a young audience while simultaneously offering viewers a glimpse of numerous fashion brands.

For another, the "Male Beauty Era" has arrived in China. According to Alibaba's Tmall, the total number of Chinese men who are buying personal-care products is growing faster than that of women (31% year-on-year in 2018 versus 29%, respectively). The total value of the personal-care items that men are purchasing on Tmall is soaring across all categories—especially for makeup, while guys are actively seeking out premium brands from all over the world (see Figure 2.4).

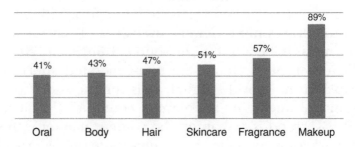

Figure 2.4   China's "Male Beauty Era"
Data Source: Alibaba Tmall.

The current trend is "shoppertainment", which means that technology is fusing the three areas—social network, online entertainment, and e-commerce shopping—together. Compared to their US counterparts, Chinese mobile apps are more advanced for content, social and commerce, and they are more transaction-based. Beyond simply opening an online shop on e-commerce platform, brand advertisers are increasingly working with KOLs (Key Opinion Leaders) at the entertainment platforms to reach potential buyers via live-streaming broadcast, which gives the followers/audiences/buyers rich product information and direct interaction with the brand messenger. For savvy mobile consumers, smart technology isn't just part of the user experience, it *is* the user experience. **(See the "Lipstick Brother Planting Weeds" box.)**

The best example of this "shoppertainment" trend is the strategic partnership between JD.com and Kuaishou formed in May 2020. JD.com, as is covered in the following chapter, is Alibaba's long-time nemesis in e-retailing. Right before JD's 2014 IPO, Tencent bought 15% of JD for US$215 million, and the two companies set up an alliance to convert WeChat user traffic at the social network into buyer traffic at JD's e-retailing platform. With the new partnership with Kuaishou, JD can potentially lean on Kuaishou videos' popularity in small towns and rural villages to advance its goal to reach hundreds of millions of up-and-coming online consumers.

Lipstick Brother Planting Weeds

Weed (*cao*) is the metaphor used by Chinese netizens to depict their ever-growing, unstoppable desire to buy, buy, buy more things. According to the online encyclopedia Baidu Baike, the weeds-to-consumerism analogy was inspired by the classic poem from the Poet *Bai Juyi*, which states "even bonfire couldn't burn it all / with the breeze of spring it grows again". (Sometimes *cao* is translated into "grass", but it seems that the term "weed" better depicts the thriving, fast growth of consumption desire than the less vigorous "grass".)

Three phrases relating to weeds describe the psychological stages of each consumer's journey (see Figure 2.5):

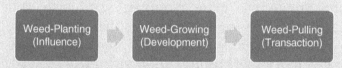

Figure 2.5   Weed – The three psychological stages of a consumer

1. Weed-planting (*Zhongcao*)—to influence someone or to be influenced to fall in love with a particular product. A person can "plant weeds" with someone else, be it friends, family or someone random online. At the same time, everyone is constantly being "Zhongcao-ed" by this person's social network posts and online reviews, comments and media articles on the internet.
2. Weed-growing (*Zhangcao*)—after someone has been Zhongcao-ed with something, he or she may continue to receive and read related information, either self-driven or internationally by internet platforms based on his or her personal profile, leading to a stronger desire to purchase or experience something.
3. Weed-pulling (*Bacao*)—the desire is finally quenched by purchasing the product or experience. (However, the term has a double meaning. It could also mean that the person has decided not to buy it anymore.)

In the context of live-streaming promotion by online celebrities, the weed-planting, weed-growing and weed-pulling essentially happens at the same time. Beauty-related products such as makeup and skin care, for example, are fertile soil for flourishing weeds. Li Jiaqi (a man, not a woman), better known as "Lipstick Brother", is an internet celebrity and the hottest lipstick salesperson.

The 27-year-old (former) shop assistant always demonstrates the lipsticks he's selling on his lips, rather than his arms, which apparently helped him to attract more than 40 million followers on the short video app Douyin, according to 2019 media stories. He even took on Alibaba Chairman Jack Ma to see who could sell the most lipstick on the Douyin platform TikTok—and won. After all, the weed business is personal.

No doubt, the mobile transformation of China's economy has also had profound implications for global stakeholders dealing with the Chinese market. Chinese e-consumers, probably among the most mobile-advanced in the world, are still demanding more innovative online shopping experiences, while companies are rushing to offer ever more sophisticated online services. In this fast-changing market, even for multinational corporations that have done business in China for many decades, a comprehensive rethink of their strategies in China may be necessary.

Meanwhile, the mobile infrastructure also opens new opportunities to foreign merchants who do not have, or need to have, a physical presence in China. To a large extent, for both multinational companies in China and any merchants thinking of bringing products to China, mastering mobile internet strategy and developing social network advertising will separate the winning retailers from the rest of the pack from their competition. For foreign companies going to China in search of their slice of the digital economy pie, the annual Singles' Day can be a good start.

Singles' Day: Hard Tech Drives the e-Shopping Festival

What is the most celebrated Chinese holiday globally by the Chinese and by everyone else? Here's a hint: It's not the Spring Festival, also known as Chinese New Year. It's November 11, known as Singles' Day.

Every November 11, billions of Chinese, at home or abroad, passionately participate in the 24-hour online shopping extravaganza. It is also a global festival, as international buyers and sellers from more than 200 countries and regions get involved. (**See the "November 11—From Singles' Celebration to Global Festival" box**).

In 2019, at the speed of the internet, US$1 billion worth of orders was placed in the first 68 seconds, scooping up everything from Apple iPhones to jewelry and even cars. The total trade volume of the day ("Gross Merchandise Value" or "GMV") was more than US$38 billion (268.4 billion RMB). By far, the November 11 festival is the world's largest online shopping day, beating Black Friday and Cyber Monday combined.

November 11—From Singles' Celebration to Global Festival

November 11 was first known as the "Guanggun Jie" ("Bare Sticks Festival" or "Bachelors' Day") in the 1990s. It was celebrated by students at Chinese universities because the numerals that form the date, 11/11, looked like four solitary stick figures. Over the years, this loosely defined holiday has become a celebration for all singles. The day has also become much more gender-inclusive by becoming "Singles' Day" that we see today.

Gradually, as the Chinese economy continued to flourish, the event started to feature shopping as an intrinsic part of the celebration. When Alibaba provided access to e-commerce through its website, Singles' Day became a virtual festival for everyone, single or married, local or a part of the diverse Chinese diaspora overseas. Alibaba launched the Singles' Day shopping festival in 2009 as a promotional event to raise awareness of the value of online shopping. Initially having just 27 merchant participants, Singles' Day in just a few years has exploded into the largest shopping day in the world.

In recent years, Alibaba has hosted an annual gala celebration titled "Double-11 Night Carnival" on the festival eve. The gala has featured celebrity appearances and performances from Pharrell Williams, Daniel Craig, Nicole Kidman, Maria Sharapova, and Adam Lambert, among others. The show is also aired on the satellite channels and streamed on China's major video sites like Youku Tudou, which was fully acquired by Alibaba in 2015 for the convergence of e-commerce and entertainment. The videos were set up to have an advertisement cross-link that enabled viewers to place orders of consumer goods that they see in the TV shows ("buying while viewing").

No doubt, the annual online shopping festival is a vivid example of the rise of the digitally connected middle class in China. It also illustrates the convergence of social, entertainment, and e-commerce on mobile platforms. However, looking beyond the shopping spree and e-commerce GMV on its surface, the Singles' Day is increasingly defined by the behind-the-scene tech developments:

First, financial services technology (fintech) for transactions. As mentioned, mobile payment is much more widely used in China than in developed markets like the United States or Europe. In 2017, the mobile payments transaction in China was already over 10 times more than that of the United States, and total mobile GMV settled through Alipay out of the total GMV already reached 90% (see Figure 2.6). Furthermore, Alibaba and its peers also provides credit lines—tiny loans—to the buyers to fund their online purchases. Alibaba's financial arm, Ant Group, has a mega microlending business called **Huabei**, which means "just spend". (JD.com has a similar payment product called Bai'tiao, which means "blank receipt".)

Many Huabei users don't have traditional credit cards and some don't qualify for bank-issued credit cards. Thanks to a proprietary credit-scoring system, Huabei can quickly approve a borrower's credit limit according to risk metrics generated by his or her repayment and related online behavioral data

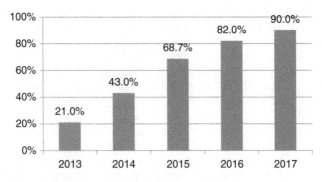

Figure 2.6   Percentage (%) of mobile GMV on Singles' Days
Data Source: Alibaba.

(number and influence of online friends, e.g.), which has helped keep default rates low. (See the detailed case study in Chapter 5.) The credit limit can be as little as a few bucks at a time to new borrowers (e.g. 50 RMB credit limit is about US$7), and Huabei has also offered temporary credit-line boosts for purchases of big-ticket items or on big shopping days. During the 2019 Singles' Day, a Kapronasia fintech analyst reportedly estimated that at least half of all purchases on Alibaba's Taobao and Tmall e-commerce platforms were made using Huabei.

Second, cloud infrastructure for processing. Processing US$38 billion transactions within 24 hours puts enormous stress on the e-commerce system, but all of this took place with "zero downtime" because Alibaba applied "multiple mature innovations to power the cloud infrastructure and other key AI-enabled features, enabling millions of businesses to reach hundreds of millions of consumers worldwide", according to a statement by Alibaba Cloud, the data intelligence backbone of Alibaba Group. At its peak, the company's Apsara Operating System supported a record 544 000 orders per second using advanced data processing and analytic capabilities. The machine translation service for Alibaba's cross-border e-commerce platform, AliExpress, was used 1.66 billion times during the shopping festival with over 200 billion words translated in 21 languages, supporting more than 100 countries and regions.

Third, smart logistics for delivery. The e-commerce industry is built on the backs of legions of package couriers—called *kuaidi*, or express delivery, to provide fast and cheap delivery services. On the busy streets of small and big cities, numerous couriers (estimated to exceed one million in total) are riding with trucks, motorbikes, scooters, three-wheeled electric carts, and any other thinkable transportation tools to make millions of deliveries every day. On each Singles' Day, the biggest headache for the e-commerce giants is logistics—the inventory, distribution, and delivery of numerous orders in a short span of time. One e-commerce company executive once pictured the challenge as "having 30 NBA basketball teams but only a few high school gyms to play".

Surely, AI came into the busy scene to provide new solutions. Alibaba's supply chain management, called Ali Smart Supply Chain (ASSC), applies AI to help online and offline merchants to *forecast* product demand. The model processes a range of historical and real-time data, including seasonal and regional variations as well as consumer preferences and behaviors. That's critical for same or next-day delivery services planning, and furthermore, it enables merchants to make quick response to shifting consumer tastes based on trends sniffed out of transactional data, allowing merchants to coordinate the flow of merchandise more efficiently—hence, reducing costs and lost sales due to out-of-stock items.

For actual delivery, Alibaba's logistics affiliate, **Cainiao**, has used AI techniques and GIS (Geographic Information System) to determine the fastest and most cost-effective delivery routes, including both rural villages and crowded urban areas. During the 2018 Singles' Day, Cainiao's total number of delivery orders processed exceeded 1 billion for the first time, yet it managed to accelerate its delivery, taking 2.6 days for the first 100 million parcel deliveries, faster than 2.8 days in the previous year.

Brushing the Sales Volume (*Shuaxiaoliang*)

The Chinese character *shua* can be literally translated into "brushing", which is why some English media used such translation to suggest people artificially "painting" a fake number for transactions. However, when *shua* is related to sales volume, it may have been taken from the term "screen refreshing" (*shua-ping*), just as a screen is refreshed and the sales volume number (*xiao-liang*) is updated when a new transaction is booked.

"Brushing the sales volume" can take many forms. The common practice is to arrange fake orders by hired part-time hands, vendor employees, or professional service providers (for the preceding reason mentioned, some media calls such people "brushers"). After the fake order is placed, the vendor still needs to arrange a delivery because the sites such as Alibaba require a unique delivery code to be entered with each order.

Initially, vendors simply sent an empty package to their contacts for a fake order; today, because Alibaba and others has increased the scrutiny of empty boxes, the vendor may send a box with clutter, such as empty bottles, piles of paper, and small rocks. Sometimes vendors may even send such boxes to strangers—those online customers whose personal information was leaked in earlier online shopping.

As the operators of online retailing platforms like Alibaba use improved techniques to monitor their websites, the vendor community also works hard to keep them in the game. Discussions on the latest best practice in brushing can be found in social media. Software that could "brush" the transaction numbers without actual purchase processing has also emerged and is readily available. A quick search on the Internet easily finds links that offer to help "brushing the sales volume", "brushing the reputation", and "brushing the satisfaction review".

And the software for brushing is also getting more sophisticated. The new versions support many setups to make the fake orders looking like human activities, such as setting up different times on the web page for different orders, different time gaps for each step of an order, and providing random reviews for transactions. Because the e-commerce players are using the online "Big Data" for AI research, "brushing" practices also means that not all transactional data is good for AI analytics. In fact, the punch line of one of those advertisements is: "Brush as many as you want"!

Finally, across all aspects of the shopping festival, Alibaba and its peer platforms are constantly collecting, gathering, analyzing, and using the user data to provide customized products and recommendations. Based on user data, the users are given different characteristic labels (e.g. "makeup lover", "sports fan", and "keen to travel"). Then, specific advertising messages are shown to potential customers based on the matching of labels. (The related personal data privacy issue is covered in Chapter 8.) The unintended consequence of the "data-driven business" is that a shadow ecosystem of fake engagement has developed, as the numbers of clicks, followers, buyers, sellers, and transactions drive vast amounts of advertising revenue.

For example, the GMV data is impacted by the market practice of "brushing the sales volume"—the fake transactions on the e-commerce site—by unscrupulous vendors. (**See the "Brushing the Sales Volume [*Shuaxiaoliang*]" box.**)

As described in the risk factors in Alibaba filings for IPO, the sellers on its site may "engage in fictitious or phantom transactions with themselves or collaborators in order to artificially inflate their own ratings on our marketplaces, reputation and search results rankings". In addition to sales volume, customer reviews and other transaction-related parameters apparently could be "brushed" as well. By all means, far from its romantic origin, the Singles' Day is all about hard tech.

BAT, TMD, and the New ATM

While the early years of China's mobile economy was mostly defined by the big three companies of Alibaba and Tencent, and Baidu (the main search engine in China)—so-called "**BAT**", a new wave of internet companies is emerging, challenging the established platforms. They are constantly developing innovations in internet search, online marketing, communications, social networking, entertainment, logistics, and other services, on both mobile devices and personal computers, to enhance users' online experience. The competition between the leading giants and the hungry up-comers is fierce.

In recent years, a new trio emerged to rival BAT. Combining the initial letters of these three companies' Pinyin words transcribed from Chinese characters, the market calls them "**TMD**". (Oddly enough, TMD also stands for the common Chinese insult "Ta Ma De", which literally means "his mother's".) What's interesting is that, although both Meituan-Dianping and Didi could trace their origin mostly back to the Silicon Valley, their business models have evolved with strong Chinese characteristics. And Toutiao has risen with pure Chinese-oriented features and services. The following is a brief introduction:

> **Toutiao**: Known as *Jinri Toutiao* in Chinese, or "Today's headlines," Toutiao is an AI/machine learning-powered content platform. Its parent company is ByteDance, which also owns the popular short video app Douyin

(and the overseas version Tik Tok). Toutiao started as a news aggregator application to provide customized content push to different audiences. Now it has evolved into a platform delivering content in a variety of formats, such as text, images, question-and-answer posts, microblogs, and, most dominantly, videos. TikTok has even become the first successful Chinese app in the mainstream United States and European markets. It was the seventh-most downloaded app of the decade, topping YouTube and Twitter, according to App Annie. According to media reports in early 2020, the company's valuation has exceeded US$100 billion, making it the most valuable media industry private start-up.

Meituan-Dianping: This is a combined company from the merger between China's leading O2O platform for local life services and customer review (equivalent to the combination of Groupon and Yelp in the United States). It's China's largest all-in-one platform for food delivery, movie tickets, restaurant reviews, and group discounts, among other things. Based on these services, it's also moving into car-hailing and bike-sharing. In 2018, Meituan-Dianping IPO-ed on the Hong Kong Stock Exchange, and in mid-2020, the company's market valuation exceeded US$130 billion, making it the third-most-valuable Chinese internet company (see Figure 2.7).

Didi: As its name suggests (meaning "honk honk" of a taxi), Didi is the largest car-hailing service in China. It achieved its dominant market position in China after a US$35 billion merger of Uber China. (Uber is a shareholder of Didi after the merger. Uber's latest annual report implies a valuation for the private Chinese company of about US53billion.) From car-hailing, it has also stepped into bike sharing and e-scooter sharing. To expand from "internet business model" to "digital tech", it has a specialized unit focusing on automated driving

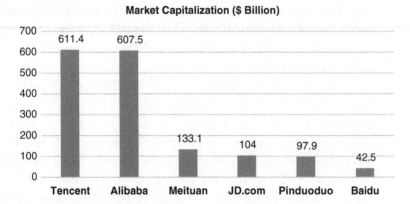

Figure 2.7   Top 5 publicly listed Chinese internet companies
Data Source: Public Markets, June 23, 2020.

research and development. In June 2020, Didi disclosed its strategic goal: to operate more than a million self-driving vehicles (so-called "robot taxis") by 2030.

Even in well-established e-commerce market, there is no sign of slowing down in innovation. One of the best examples is the recent rise of Pinduoduo, China's fastest growing e-commerce app in history. With its new model of marrying social group connections with e-commerce, its reach into more remote, poorer communities, its quick uptake with users, and its IPO within three years of its start, Pinduoduo has caught the older-generation businesses in e-commerce such as Alibaba and JD.com by surprise. (In June 2020, both JD and Pinduoduo reached US$100 billion market capitalization.) Compared to Taobao and JD.com, Pinduoduo's twist lies in its integration of social components into the traditional online shopping process, which the company describes as the "team purchase" (or "pin") model. (See the detailed case study in the following chapter.)

Meanwhile, Baidu, China's biggest search engine known as the "Google of China", has suffered significant loss as it struggles with the changing internet usage patterns. Baidu used to have Google-like predominance in China (it amassed a 70%

market share after Google exited China in 2010), but the company hasn't been as adept at developing services for the mobile age as Alibaba and Tencent. In recent years, its desktop search businesses continue losing users to smartphone. Baidu's value as the starting point of info search is being challenged as online contents become even more segregated.

In the mobile world, the customers' search is more fragmented. Instead of being glued to a PC screen and using a search engine like Baidu, people tend to use multiple channels to seek information when they move around with their mobile devices. For example, they may use Alibaba's Taobao to check out imported seafood, or Tencent's WeChat to ask friends for new movie information, or Meituan-Dianping's customer review app for promotion deals from nearby restaurants. All three have evolved into self-contained, super-app ecosystems, such that people can watch videos, shop online, and order takeout food without having to use a traditional search engine. As a result, Baidu is losing advertisement wallet share in the super competitive digital media space, and the latest rise of short video platforms like Douyin and Kuaishou does not help.

At present, Baidu is betting on new technologies for longer term growth. Having missed out on the social, mobile, and e-commerce waves of the past few years, Baidu is trying not to repeat the same mistake by going all in on AI. It has invested billions of dollars into "intelligence first", hoping its AI strategy would help the company reinforce its core search + feed business with improved monetization of users. But even that is not a sure path to success, as its heavy AI investments put pressure on its cash flow, and its self-driving projects are far from yielding profits.

Hence, the new acronym **ATM,** which replaces Baidu with Meituan-Dianping, has been catching on lately. Among the top 5 publicly listed Chinese internet companies in Figure 2.7, Alibaba and Tencent, with more than US$600 billion market capitalization in June 2020, are the two leading companies, way ahead of the rest. Meituan is the third most valuable Chinese internet company. The market probably likes the new acronym even better, because ATM also has the benefit of standing for automated teller machine, which rhymes with the rapid stock price gains of the trio.

The dust has not settled, of course. Taking a closer look at the top 5 list reveals the shocking fact that Alibaba is the only one out of the Tencent empire. For Meituan-Dianping, JD.com, and Pinduoduo—the direct competitors of Alibaba—Tencent is not only a strategic partner, (the omnipresent WeChat app provides approximately 1 billion monthly active user traffic for their e-commerce and online service businesses), but also a strategic investor. Not widely known by the public, Tencent holds 18.1% of Meituan-Dianping, 17.9% of JD.com, and 16.5% Pinduoduo, according to companies' public filings in 2020 (see Figure 2.8).

Going forward, the partnership, competition, and cross-investments among the established internet companies—think of BAT, TMD, or ATM—and upcoming younger players will make the mobile market extremely active and dynamic. The Chinese market is poised to be a trend-setter, rather than a trend-follower, in next-generation mobile devices and services. As Part two shows, Chinese tech companies are looking beyond the mobile internet and invest into iABCD applications for an edge in the next phase of competition, making China one of the most interesting centers of innovation in the world.

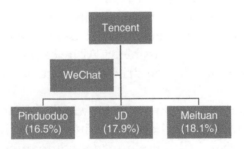

Figure 2.8   The bigger-than-you-thought Tencent empire
Data Source: Public Filings of Companies, 2020.

PART II

China's Digital Transformation and Innovation

The digital transformation of Chinese companies is reshaping retail, financial, entertainment, transportation, lifestyle services, education, smart hardware, and many traditional industries. China, one of the most interesting innovation centers in the world, is where the new generation startups are pioneering 5G iABCD innovation.

CHAPTER

3

Big Data on the Digital Middle Class

- The 4Cs Model: Know Your Customer
- Freshippo OMO: E-commerce Will Die
- C2M—Pinduoduo and Social E-commerce
- Bullet Screens: Hollywood Losing the "Big Data" Movie Battle
- Post-pandemic Consumers Come of Age

The 4Cs Model: Know Your Customer

What is the single most important and difficult issue that Walmart, Procter & Gamble, and Unilever are facing in the Chinese market? The answer is very simple: *they no longer know Chinese consumers as well as before.* This is because the digitally enabled Chinese consumers are much more sophisticated than they were years—let alone decades—ago.

As mentioned in the previous chapter, China's digitally connected new middle class has led to a seismic change in the Chinese consumer market. Not only is this new group of consumers extremely comfortable with mobile e-commerce, but they also have enjoyed unprecedented exposure and access to foreign brands, with social networks having tremendous influence on their purchasing decisions. As a result, they have high expectations both of quality and of service that did not previously exist in China.

Furthermore, the rise of the digital middle class and its spending power is occurring with a generational shift, as the younger consumers in China (generations of the 1990s and 2000s) have become one of the fastest-growing and increasingly influential segments of Chinese consumers. In the next 10 years, about 200 million people born in the 1990s will be starting families, and about 150 million in the 2000s will transition from school or university to the Chinese workforce.

Unlike their parents, these groups of the 1990s and 2000s grew up in a China marked by unforeseen wealth, exposure, and technology. The younger generation consumers enjoy the status conferred by luxury brands, and they routinely trade up to the next tier of these items. White collar professionals constantly look for the newest products and trends from the foreign movies and TV shows on the video streaming sites. Even young migrant workers, who are clearly in the lower income bucket, may spend a month's wages on high-end products like Apple iPhones.

Because young Chinese consumers are both voracious e-shoppers and active social network users, a new social phenomenon, *shai*, has emerged. In Chinese, *shai* means "to put something under the sunshine". In social media, the term describes the young generation "showing off" their lifestyle, such as a luxury brand recently purchased or a fancy restaurant currently being patronized. As a result of this phenomenon, young Chinese consumers have a strong demand for customized and personalized products and experiences.

In fact, in another sign of their growing sophistication and confidence, young Chinese are increasingly open to local brands, whereas Chinese users traditionally trusted well-known international brands a lot more than domestic ones. Western brands, therefore, cannot simply turn their global advertisement into Chinese language as they did during their entries into the China market in the 1980s or 1990s; instead, they need to rethink their Chinese advertising campaigns within this new context. They must constantly create new aspirations and identities in their stories to keep the young Chinese consumers engaged.

For example, according to a 2016 survey of the Chinese market on popular shampoo brands, conducted by *CBNweekly*, consumers tend to buy shampoo because of either best product performance or unique "selling points". While P&G brands had dominated the market for many years, by 2016, only Head & Shoulders, its most functional brand, remained In the top 10 list. Its Rejoice and Pantene brands were pushed down by the more high-end brand Schwarzkopf. The British brand Lush made it to the list because it attracted large number of followers by promoting its products' handmade, natural, fun, and edible features; the Korean brand Ryo won lots of female customers' hearts by connecting its brand image with Korean romantic TV shows sweeping across China on video-streaming apps (see Figure 3.1).

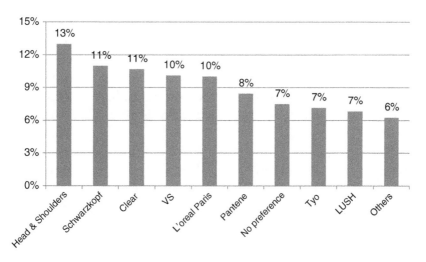

Figure 3.1   Top 10 shampoo brands in China
Data Source: *CBNweekly* survey, 2016.

To make the matter more challenging to major corporations and established brands, new digital technologies are creating even more information channels for customers, and at the same time, enabling digital platforms to collect more consumer data than the brands themselves:

- Machine learning (AI) and virtual assistants will enable the provision of instant push notifications and customized information for consumers.
- When mass adoption of VR technologies occurs, consumers will have three-dimensional information and perspectives on consumer goods and services (providing businesses with more consumer behavior data).
- Big Data analysis will allow retail channels and platforms (not necessarily the brands or service providers themselves) to have a richer, deeper, and more accurate understanding of their target audiences and be better able to predict their behavior.

Therefore, all businesses must rethink their consumer-engagement strategy fundamentally. Bland, mass market products shipped from standardized assembly lines are losing their glamour, and the young "experiential generation" further demands "personalized" shopping as it seeks more businesses to provide richer consumer experience. In the coming years, the key to all industry sectors is to personalize products and services to serve the varying needs and demands of millions of Chinese consumers.

The solution is to put the customer at the center of business operations (see Figure 3.2). To do that, corporations have to make significant investments into the Big Data technology to develop a deep and rich understanding of their customers—not only about their retailing pattern, but also about their overall spending habits as well as their social network and behavior. To win over repeating customers, merchants must deliver a distinguishable and personalized shopping experience by

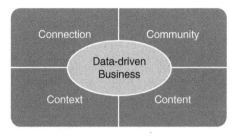

Figure 3.2   The 4Cs in the smart economy

understanding not just what customers want and need, but also where and how they want to experience it (e.g. the integration of entertainment contents with shopping activities).

In short, there is no longer a single, one-size-fits-all definition of the Chinese consumer. The new Chinese consumers have grown a lot more sophisticated and diverse than before, and their data in the digital age has truly become "Big Data". In the following sections, the successful business models in the examples of Freshippo "new retail", Pinduoduo "social e-commerce", and "internet+" movies are completely data-centered, seamlessly integrating consumer engagement around the 4Cs: Context, Connection, Community, and Content.

Freshippo OMO: E-commerce Will Die

In 2015, when China's e-commerce was powering forward at a rapid pace, Alibaba founder Jack Ma shocked the market by predicting, "e-commerce will die." Of course today, Alibaba still dominates e-commerce with two powerful marketplaces: Taobao (C2C), an online bazaar that offers a huge range of consumer goods, and Tmall (B2C), a more refined platform for established brands (see Table 3.1). However, Tmall is more than a pure distribution platform and has a strong value proposition for what Ma calls "new retail": an enabler for brands and merchants to reach new customers and service repeat customers through marketing tools and consumer data insights.

Table 3.1   Alibaba's Taobao and Tmall marketplaces

	Taobao	Tmall
Year of start	2003	2008
Concept	C2C bazaar (comparable to eBay)	B2C mall
Setting	Sellers post new and used goods for sale or resale	Each brand can set up its own virtual store in the mall
Users	Individuals, small merchants	Popular brands from home and abroad, such as Apple, BMW, and Tesla
Cost to users	No commission fee; main revenue from online marketing services for vendors	Sellers pay a deposit to list on the site, and Tmall earns commission on transactions

At the center of "new retail" is the concept to offer customers a seamless integration of online/offline shopping experiences. In today's hyper connected world, a consumer's buying journey spans across offline and online. Merchants must make sure they serve the consumers using all possible channels because consumers are getting smarter every day with the increasing penetration of smart devices and can easily handle all channels. The overall trend is that the online and offline consumer activities are more than ever integrated, and the line separating the two is increasingly blurred.

The model of seamless integration of online/offline promises consumers unprecedented choice, convenience, and simplicity, but with hyper-connectivity, the manufacturers and vendors are also facing exploding complexity and they have to make significant investments into the Big Data technology. Using data analytics, merchants could potentially create a consistent and smooth shopping experience across different channels—physical stores (offline), company websites (online), and mobile commerce (linking online and offline).

Not surprisingly, more data is being collected offline as most people still spend more time offline than online, and the data from offline spending scenes have more dimensions and depths. **To that end, "new retail" first started with e-commerce giants making acquisitions into the offline physical stores**

Figure 3.3   E-commerce giants acquiring offline stores

(years before US e-commerce giant Amazon bought the offline supermarket chain Whole Foods Market, see Figure 3.3).

As early as August 2015, Alibaba invested US$4.6 billion to become the second largest shareholder of Suning Commerce Group, the nation's largest home appliance retailer. (Alibaba's rival in e-commerce, JD.com, also entered into physical store retailing by investing US$700 million for a 10% stake in Yonghui Superstores, which was China's fourth-largest supermarket chain). For Alibaba (and JD), the goal was to leverage the offline retailers' existing networks of brick-and-mortar stores to boost its supply chain and diversify its offline offerings (and subsequently, accumulate more data).

Furthermore, backed by its mobile payment and digital technology, Alibaba (and its rival Tencent) also set up unmanned convenience stores for more offline traffic and user data. To enter those unmanned convenience stores, customers need to scan a QR code to open the door. Because the assorted goods in the store all have embedded radio-frequency identification (RFID) chips, when patrons select their favorite items and leave the store, the storefront scanner reads the RFID chip and sends the bill to the customer's mobile payment app. The whole process involves computer vision, biometric recognition, and sensing technologies to track customers and their purchases, resulting in automatic consumer payment at the checkout gate.

The latest and most interesting development of "new retail" is what can be described as "digitalized offline stores":

that is, digital technology-enabled offline stores that provide both a mobile online setting and traditional offline shopping experience—in one and the same location. The Freshippo (known as *Hema* in Chinese) stores of Alibaba are the perfect example. On the one hand, they are well equipped with digitalized operating systems, in-store technology, supply chain systems, consumer insights, and a mobile ecosystem. On the other hand, they combine the features of offline supermarket, restaurant, and direct-sale fresh market of vegetable, fruit, and seafood. Together, they provide a seamless shopping experience for consumers that merges online and offline activities (known as "OMO"; see Figure 3.4).

At the center of the exhibition areas of Freshippo stores are fresh vegetables, fruit, and seafood (Freshippo partners with Tmall, Alibaba's e-commerce platform, for direct and cost-effective products sourced from its original production). As the young middle class pursues a healthy lifestyle, fresh food is a new kind of "affordable luxury" that appeals to them. Shoppers can select live seafood and send it to the kitchen to be cooked and eaten on the spot. Or they can use the app to order their groceries for delivery. Since each store serves as a warehouse and its own fulfillment center, Freshippo's proprietary fulfillment system promises to deliver within 30 minutes to consumers who live within a three-kilometer radius.

The Freshippo model is completely data-centered. The stores supply product information and collect consumer data in multiple, interactive ways:

Figure 3.4   OMO seamlessly integrating online and offline

First, information about the food is provided on the mobile terminal. In recent years, Chinese customers' faith in local food has been shaken by a slew of food safety scandals, which appear to have turned everyone into a food safety expert. (**See the "Smart Chopsticks and Blockchain Chickens" box.**) Freshippo offers a mobile app that allows consumers to search for products and place orders while browsing in the store. The shoppers can scan the barcode on all items in Freshippo to trace the products' origin, delivery, and nutritional information. Furthermore, such information helps young shoppers to become connoisseurs of the exotic, so that they could excitingly show their lifestyle elevation by posting pictures of their adventure with luxury food on social media platforms.

Smart Chopsticks and Blockchain Chickens

Because of the widespread food safety scandals, Chinese consumers' confidence in food, especially when it's domestically produced, is low. In 2008, milk powder tainted with melamine, a toxic industrial compound, made 300 000 babies ill—six died. Since then, most Chinese parents have turned away from locally produced brands, and the supermarkets in Australia, Japan, and Hong Kong often saw Chinese tourists buying up baby formula on their shelves. (In May 2020, the milk scare came back to Hunan Province, where toddlers were found to develop rickets and a skull deformity resembling "big head" after being raised on a protein drink sold to their parents as baby formula.)

More stomach-churning food safety scandals followed the deadly tainted milk case. Chinese consumers have also encountered watermelons that exploded from the misuse of a growth accelerator chemical, lamb made of rat meat, pork soaked in a detergent additive, and cooking oil recycled from waste oil collected from restaurant fryers, grease traps, or even sewer drains (known as the "gutter oil"). Sometimes it is just hard to figure out how far away the food is from "fresh". In 2015, Chinese authorities seized 3 billion RMB (close to US$500 million) worth of frozen beef, pork, and chickens that dated as far back as the 1970s. "A bottle of 1982 Lafite plus a piece of 1970s steak and a pair of 1980s chicken wings," wrote one Chinese user on the Sina Weibo microblog (the Chinese equivalent of Twitter), adding "Bon appétit!"

On April Fools' Day 2014, China's search engine giant Baidu offered a video clip on smart chopsticks that could determine whether a dish contained gutter oil. According to Baidu, when the video was made, it had no serious intention to pursue it as a product. However, because the fake advertisement generated so much buzz on the social media, Baidu decided it could be a timely innovation.

At the company's annual technology conference in September 2014, Baidu's Chief Executive Robin Li unveiled the "smart chopstick" prototype that was called Baidu Kuaisou. The utensils were equipped with sensors to collect data on pH levels, peroxide value, and temperature, and they could be connected to a smartphone app to provide users with analyzed readings on the oil being tested. But some food expert immediately warned that gutter oil producers could outsmart the smart chopsticks. Because the sensors could only take a small number of variables for its analysis, the gutter oil producers could, according to the experts, easily add relevant chemicals to give its oil products a false safe reading.

Now the cutting-edge technology for food security is blockchain. It is used to collect data about the origin, safety, and authenticity of food, and provide real-time traceability throughout the supply chain. This has traditionally been challenging due to complex and fragmented data-sharing systems that are often paper-based and can be error-prone. The blockchain solution would provide consumers and regulators far more information about the food on the shelves: its source and region, its shipping process, and its inspection and certification, among other information.

For example, JD.com, a major e-commerce competitor of Alibaba, piloted a blockchain application for consumers in select Chinese cities to track meat from Chinese beef producer Kerchin based in Inner Mongolia. Working with the beef producer, JD allowed consumers to access detailed information, such as the cow's breed, when it was slaughtered, and what bacteria testing it went through. Furthermore, it worked as Australian exporter InterAgri used blockchain to track the production and delivery of Black Angus beef from import.

Tech firms nevertheless thought they could go even further with blockchain. In 2019, GoGo Chicken, a poultry monitoring technology based on blockchain, was developed by ZhongAn Technology, a subsidiary of the Chinese online insurer ZhongAn Online, to chronicle chickens' life stories to prove whether they are organic or not.

According to the company, each chicken would wear a tracking device on its foot, which automatically would upload its real-time movements through the supply chain to the blockchain database. Sensors would monitor temperature, humidity, and other aspects of each chicken's environment, while algorithms would evaluate the bird's health using video analysis. ZhongAn plans to roll out this technology to hundreds of Chinese farms by 2020, and is confident that eco-conscious consumers will be happy to pay a premium to ensure that the chickens they buy are truly cage-free.

Second, offline interaction supplements online information. Unlike clothing and shoes, fresh food—for example, seafood—is not a standardized product. It is difficult for consumers to gauge the quality of products purely from information provided online. Offline physical stores enable them to interact with representatives from traditional outlets to get to know the products better. Middle-aged shoppers, in particular, have more trust in traditional stores where they can peruse seafood directly. They prefer to first purchase imported seafood at physical stores before ordering the same products online in the future.

Third, shoppers can make "product trial" in small amounts. Before shoppers commit to purchasing, they can "try and pay" along the way, just like they do with clothing, using retail shops in malls as showrooms or fitting rooms. For example, if a customer receives the latest promotion of Canadian blackberries from the Freshippo app, she can buy one small box using the Freshippo app and Alipay (Alibaba's mobile payment); before piling them up in a shopping cart, she can taste them in the leisure area and then buy a dozen more if they satisfy her expectations (or order the dozen online for delivery later).

Fourth, Alipay is the only accepted payment method for checkout. The mobile-payment terminal of Alibaba (see the details in Chapter 5 in connection with fintech discussions) records consumer data for both online and offline transactions. All in all, the above consumer activities are fully captured digitally. Their historical data enables the Freshippo app to generate highly personalized recommendations and targeted promotion, enhancing the shopping experience for consumers when they shop online or offline later.

Finally, the entire Freshippo store operation is digitalized in the background, including in-store technology, digitized inventory, and supply chain systems (sourcing, sales, and delivery). Through data technologies, Freshippo shortens the sourcing process and increases supply chain transparency and visibility. For the consumers, Freshippo uses transaction data to personalize recommendations and geographic data to help plan the most efficient delivery routes.

Just as predicted at the beginning of this chapter on the Big Data trend, the Freshippo stores assume huge upfront costs for data technology investments. What is promising is that the Freshippo stores' economy of scale is expected, and its partnership with Tmall brings savings from sourcing (through direct sourcing and bargaining power). By the end of 2019, Freshippo has opened 197 stores in over 20 major cities across China (see Figure 3.5), of which online sales account for more than half of total store sales, showing the "offline" stores have strong "online" characteristics.

As illustrated in the Freshippo case, to provide customers seamless experience, the digital stores of internet firms use both online and offline touchpoints to cover the full decision and purchase journey of consumers—searching and comparing products online, trying them out at offline shops, paying online or at offline outlets, arranging delivery online, or picking up the products from physical shops. Storefronts at convenient locations—like the Freshippo stores—attract foot traffic as do traditional retailers, but they also have an "online persona" managing marketing, logistics, post-sale services, and customer engagement (see Figure 3.6). As a result, the

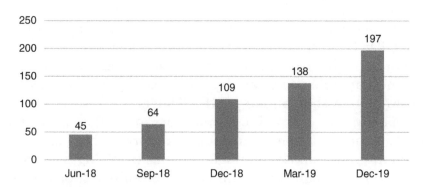

Figure 3.5   Rapid expansion of Freshippo stores
Data Source: Alibaba.

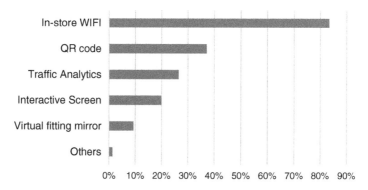

Figure 3.6   High percentage of supermarkets adding online persona
Data Source: Ministry of Commerce, 2017–2018 China Supermarket Development Report.

difference between online commerce and offline commerce is disappearing.

In summary, the "new retail" in Alibaba's term is essentially OMO (online-merge-offline, see Figure 3.4). To be successful, it is as important for e-commerce companies to associate with physical stores as it is for bricks-and-mortar retailers to have an online presence. By now, Alibiba founder Jack Ma's surprising statement "e-commerce will die" becomes easy to understand. The future e-commerce model—if the term *e-commerce* is still viable—is a new retail platform that seamlessly links customers across multiple screens of digital devices, providing standardized products as well as specialized goods and experience offerings, connecting online contents with offline activities.

Meanwhile, it's important to highlight that the Freshippo retail chain is primarily located in tier 1 and tier 2 cities, where the average spending power of consumers is the highest in China. According to the company's filings, this new business will expand by increasing Freshippo's store density in existing cities in order to improve consumer coverage and delivery efficiency. As the Pinduoduo case in the next section shows, China is a vast and diverse market, and Freshippo only represents a small segment, and there are many ways for Chinese consumers to seek "consumption upgrade".

C2M: Pinduoduo and Social E-commerce

Improved living standards and growing wealth generally drive demand for products with higher quality or safety standards—the so-called "consumption upgrade" (*xiaofei shenji* in Chinese). Within China, this dynamic has materialized as a trend of premiumization: Alibaba's Freshippo is the answer to the middle class's quest for fresh and safe food, and **JD.com**, Alibaba's main rival in digital retailing, has its own offerings.

By many measures, JD.com is not nearly as big as Alibaba, especially when it comes to gross merchandise volume (GMV), that is, the total value of all online transactions. That difference, however, has a lot to do with the two companyies' models (see Table 3.2). As mentioned, Alibaba is mostly a platform provider for buyers and sellers (like the combination of eBay and PayPal in this aspect), while JD.com buys inventories and sells them to consumers as a direct online retailer (more like Amazon.com). In recent years, their competition is much more direct and fierce than the gap of GMV sizes suggests.

Since 2007, JD has been the first and probably the only Chinese e-retailing company to invest substantially in its own logistics network. JD's rationale is to handle the whole process

Table 3.2   Main differences between Alibaba and JD

	Alibaba	JD
Business model	Online platform for retailers	Online direct sales
Business lines	E-commerce marketplace. Broad, various services; integrated online ecosystem	Mostly direct retailing and logistics services; Into related e-retailing finance
Transaction volume	Huge	Relatively smaller
Logistics	Partnerships with third-party companies to provide logistic services	Owns nation-wide warehouses, dispatch centers, and delivery networks
Source of profits	Marketing and services income; commissions	Difference between retail price and cost
Profit margin	High	Low

itself from online orders to physical drop-off to better control the delivery service and product quality. Also, JD's supporters believe this model sets JD apart from its competitors by having fewer counterfeits and lower chance of damage during deliveries (of particular importance in the case of healthcare products or fresh fruit, e.g.).

For example, during a recent "Singles' Day" sales season, JD.com used an advertisement slogan "same price, buy genuine". There was also an advertisement in which a young professional got embarrassed during a job interview when her new red dress bought online left marks on the interviewer's couch. No name was mentioned in those ads, but it seemed everyone knew which online firm was being referred to. Alibaba's founder Jack Ma, however, argued a few years ago that JD.com's model would be difficult to scale up to deal with the explosive growth of China's e-commerce. (**See the "JD Will Become a Tragedy?" box.**)

JD Will Become a Tragedy?

"JD.com will become a tragedy. I have warned everyone of JD.com's model from the very first day".

That statement was apparently part of a private conversion between Jack Ma, the founder and chairman of Alibaba, and a personal friend. In the private conversation, Ma criticized the business model of its e-commerce arch-rival JD.com, but without his knowledge, his friend incorporated it in a new book. Thus, Ma's colorful commentary got publicized in early 2015, which triggered immediate response from JD's Founder Richard Liu, creating a drama for billions of Chinese people during the New Year's holiday season.

"How many people does JD.com hire right now?" Ma asked and then answered the question himself: "50 000 people!" Alibaba has expanded slowly [implying a longer history of existence], and by now we only have 23 000 employees. Why do I choose not to do the delivery myself at Alibaba? Right now, JD has 50000 employees, adding on between 30,000 to 40,000 people for inventory, but they only handle 2 million parcels a day. But Alibaba deals with 27 million packages a day. How would you deal with that?

"In ten years, China's e-commerce markets may have to deal 300 million packages a day. You may need to hire 1 million people; how can you manage that"? By suggesting that a self-run delivery system was impractical given the market size, Ma concluded that the JD business model was a "tragedy". "It's not that we are better", he said, "It's an issue of direction". Hence, his advice: "Never ever touch JD.com".

In response to Mr. Ma's direct challenge to its business model, JD chose to deliver its elaborate response in a Chinese poem, titled "We Will Do Our Best and Time Will Prove It". The JD poem sarcastically praised Ma as a lonely wise man, on whose "sagacity" JD would meditate.

Then the poem went on to make a comparison between the two companies, suggesting JD was focusing on improving customer services, while Alibaba was short on quality control when managing a large marketplace. As the poem put it: Alibaba had "a grand posture at podiums", while JD bowed down to "serve at store counters"; Alibaba was good at "glamorous talking", while JD "built on infrastructure in hard labor"; Alibaba as a marketplace "earns money easily", but JD chooses to internalize inventory and delivery to "have better quality control", in hope of "win[ning] over customers". In summary, even though "we (the two companies) look at each other often, we are on different paths. You are minding our business so much that we are moved to tears".

The war of words between the two internet giants was quite a drama, and the strong personalities of the two founders added more color to the exchange. However, the e-commerce industry looked at that as a serious debate on which model represented the winning e-commerce model in the long term. On Ma's challenge to JD's sustainable growth, JD's response did not directly provide an economic analysis to justify its own model. To some extent, in JD's own poem, it indirectly admitted that its model involves high cost and input ("hard labor"), but does not necessarily generate as high a return as Alibaba (the latter "earns money easily").

To some extent, the rapid rise of Pinduoduo in recent years, which now claims itself as the second largest e-commerce platform in China, adds evidence to Jack Ma's prediction. Founded in September 2015, Pinduoduo, is an online group-buying site that's more like the Alibaba model than that of JD.com. Pinduoduo also acts as a marketplace, and its twist is in its integration of social components into the traditional online shopping process, which the company describes as the "team purchase" (or "pin") model. With its new model of marrying social group connections with e-commerce, its reach into more remote, poorer communities, its quick uptake with users, and its IPO

Figure 3.7   Together, more savings, more fun

within three years of its start (PDD is its ticker symbol), Pinduoduo has caught the established businesses in e-commerce such as Alibaba and JD.com by surprise.

Unlike its two main rivals, Pinduoduo tells the market that for a large group of users, consumption upgrade still means buying stuff cheap in a mobile, fun way. Corresponding to the company's names in three Chinese characters, Pinduoduo's value proposition is "Together, More savings, More fun" (see Figure 3.7). Pinduoduo users, by sharing the product info on social networks like WeChat (Tencent became an investor of Pinduoduo in 2016 and provided a strategic channel for Pinduoduo to reach the billion-user traffic on the WeChat network) can invite their contact to form a shopping group to get a lower price for their purchase. This social e-commerce mechanism makes the shopping experience more interactive and the users more motivated.

On Pinduoduo's platform, a buyer can initiate a team purchase and share product information on WeChat to invite his or her social contacts to form a shopping team. (Alternatively, a buyer can choose to join an active team purchase listed on the platform, which is initiated by other buyers who may or may not be his or her social contacts.) Would-be buyers have 24 hours to complete a buying group, and the app shows the discount increasing with each additional buyer added. For shoppers who love a bargain, this creates urgency and excitement. For their friends, it can be difficult to turn down an offer to help them save money. Psychologically, it's an irresistible experience. (If the minimum team size is not reached within 24 hours, the team purchase order may be cancelled with all payments made by the buyers refunded.)

Because of its group purchase format, Pinduoduo is some-time described as "Groupon on Steroids". But its founder, Colin Huang, knows his company's model better. Where Groupon focused on "want to have" goods and services like restaurants, massages, and travels, Pinduoduo offers deep discounts for bulk purchases of everyday "need to have" household items. Meanwhile, it gamified shopping for the most mundane everyday products to drive user engagement and retention. As such, Huang describes the company as a combination of discount retailer Costco and entertainment property Disneyland.

By mid-2020, the 5-year-old Pinduoduo's US$100 billion market cap is the equal of JD.com, and its 536 million annual active buyers (2019) is second in China only to Alibaba. Of course, the larger number of active buyers, order numbers, and monthly active users (MAU) of Pinduoduo is partly attributable to the different business models (in the same way as Alibaba versus JD; see Table 3.2). However, it is remarkable that the youngest player has proved to be best at user stickiness—its users come by more often and spend more time on its site. According to a February 2020 report by the Sinolink Securities, Pinduoduo's user growth is the fastest among the three, and the average DAT (daily active time) on the Pinduoduo platform is the longest (see Figure 3.8).

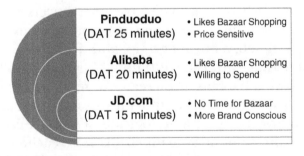

Figure 3.8   Different platform users' profiles and DATs
Data Source: Sinolink Securities, 2020.

The "user stickiness" is the key to Pinduoduo's success, which is based on accurately segmenting Chinese online users' demography:

First, Pinduoduo's main target market is price-conscious buyers in tier 3 and below cities and in China's rural areas (almost half of China's population is still in rural areas)—an underserved group. China's mobile economy is increasingly defined by the demands of the "small town youths", a term loosely covering the young netizens in markets other than the tier 1 cities. As the major city markets have become more saturated, the rural areas provide the most growth potential for the e-commerce companies. Although the small-town youths may have a preference for cheap goods, they are more frequent buyers, spend more time on e-commerce platforms, and collectively, have remarkable purchasing power.

Second, because its main customers have "more time than money", Pinduoduo designs its "team purchase" model to transform online shopping into a dynamic social experience that mirrors the social interactions consumers tend to have offline. Supported by the WeChat social network, it has consciously built the platform to resemble a "virtual bazaar" where buyers browse and explore a full spectrum of products while interacting with one another. Furthermore, Pinduoduo continues to add fun and social elements to its platform. The Duo Duo Orchard, an in-app game launched in 2018, is a good example.

In Duo Duo Orchard, users may choose a virtual fruit tree (the choices vary depending on seasonality) and then "water" it to grow from a sapling to a fully-grown fruit tree, when they will win a prize—a free box of (real) fresh fruit from Pinduoduo. To earn "virtual water" for their tree, players need to fulfil a variety of missions such as shopping on Pinduoduo's platform, sharing products, and inviting friends to join the platform. Players can also check on how their friends' trees are doing—either to help them water their trees or steal their water droplets. The game is not only effective to incentivize consumers to browse and purchase, but also to motivate users to interact and enjoy shopping with friends. In 2019, the company saw over 11 million daily active users (DAUs) logging on to the game to play and explore.

Third, as users interact more on the platform, the company is able to gain insights on their shopping preferences and needs, and in turn, use that to offer more relevant products to them, driving a virtuous cycle of even greater user satisfaction and engagement. According to the company's 2019 annual report, it has built up Big Data analytics capability that can efficiently handle complex computing tasks of billions of data instances and millions of analytical dimensions. For example, for one transaction, Pinduoduo not only looks into the basic order information, but also the related buyer behavioral data, such as how long the buyer spent on browsing and reviewing a particular product and products of similar categories. By aggregating users' demand based on their preferences, the platform can help users realize more savings and do so in a fun and engaging manner.

In summary, Pinduoduo moves e-commerce away from concentrated searches to social interactions. In contrast to the conventional search-based "inventory index" model, this social e-commerce model tries to bring out fun and excitement of discovery and shopping, which fosters a highly engaged user base—and extends user time on the platform. The established e-commerce function is reversed: now it's "Goods find People" rather than "People find Goods" (see Table 3.3).

Therefore, the Big Data generated from the "pin" model is not only an efficient tool for user engagement and expansion, but also an opportunity to improve the supply chain efficiency of the retail market. Traditionally, developing a new product often involved a lengthy process of market research, focus group testing, feedback through distribution channels and then large-scale production. By contrast, Pinduoduo has developed the **"C2M"** (Consumer-to-Manufacturer) model, whereby

Table 3.3   Social e-commerce versus traditional e-commerce

	Traditional e-commerce	Pinduoduo's social e-commerce
Model	Search-based shopping	Fee-based shopping
Business Philosophy	"You know what you want to buy"	"You don't know what you want, but are happy to discover"
Platform Role	Acts like a super brain that can answer all users' questions	Acts like a personal agent that can provide users with tailored advice
Platform Setup	Platform hosts an index of all SKUs and brands	Platform provides better dynamic recommendations through interactive features and AI capabilities that learn user preferences
User Experience	Solitary shopping experience	Fun and interactive shopping experience
Transaction Flow	Design --> Manufacturer --> Consumer	Consumer --> Design --> Manufacturer

the aggregated user interests and demand, as demonstrated by the "team purchase" activities, could be relayed to merchants, so that they can adjust their production and sales plans accordingly. As such, factories can compress the new product development process and reduce costs.

Take agricultural products, for example. For historical reasons, China has relatively less arable land per capita, which is different from countries like the United States, where large-scale farms are prevalent, and the production and transportation of agricultural products could be highly industrialized. Thanks to its dynamic AI engine, Pinduoduo can develop robust user profiles to provide market demand information that was previously inaccessible to farmers. The farmers can also use its platform to aggregate larger volume orders for their products. The large demand helps the farmers to be less dependent on distributors and makes it possible for them to sell directly to consumers. As a result, consumers can acquire fresher and safer products for a lower price, while farmers earn more, thanks to lower distribution costs and larger orders.

Diao-si ("Grassroots") versus Gao-da-shang ("High Society")

The term *diao-si* involves a vulgar reference to the male genitalia, and it became a popular term in the internet era as a synonym for "loser". The term was originally one for the young men who lived on the margin in an otherwise booming economy. They were in the workforce after school education, but were paid so poorly that they could not even have a girlfriend. Nowadays, the term is no longer derogatory and broadly applied, sometimes self-jokingly, to anyone who works at an entry-level position, has not purchased a home, or has little purchasing power.

In contrast with *diao-si, gao-da-shang* is an internet buzz word linked to high society and chic lifestyle. The term is a combination of three words, including *gao duan* ("high end"), *da qi* ("elegant" or "regal"), and *shang dangci* ("high grade"). It is often used to flatter, with a touch of humor and sometimes sarcasm, a luxury product or chic lifestyle. For Chinese smartphone brands, they have to offer a wide range of products to cater to the different purchasing powers of *diao-si* and *gao-da-shang* consumers.

When a brand identity was formerly linked with high quality and low price, the risk it faced is that some of its users may use it as a transition product before they could afford a luxury brand. When those customers have a greater surplus of purchasing power, they may not continue purchasing their desired products from digital platforms known for cheap items. A brand must evolve with its consumer base's growing purchasing power because deep in the hearts of many *diao-si*, they aspire to join the *gao-da-shang* class someday.

Going forward, Pinduoduo and Alibaba are set to have fierce and direct battles. Both are striving to cover a customer spectrum as broad as possible, from price sensitive small-town youths to brand-conscious cosmopolitan white-collar professionals. (**See "*Diao-si* ['Grassroots'] versus *Gao-da-shang* ['High Society']" box.**) On the one hand, Alibaba is also focusing more on lower tier city markets and rural areas. Since 2019 it has assigned more resources into its flash sales platform *Juhuasuan*, which provides heavily discounted group-buying sales in direct competition with Pinduoduo.

On the other hand, Pinduoduo is also expanding into more high-end products, aiming to acquire more "rich" customers. (According to the 2020 report by the Sinolink Securities, in terms of the average annual spending amount, the spending of customers on Alibaba's sites is 1.6 times of JD customers and 5.6 times of Pinduoduo customers.) It has launched a "new brand initiative" campaign that aims to support 1000 manufacturers in developing their own brands through the C2M model. Just like the argument between Alibaba and JD.com, time has to be the perfect judge and fair mediator on this fight.

Bullet Screens: Hollywood Losing the "Big Data" Movie Battle

For 2019, total ticket sales at Chinese cinemas reached a new all-time high of US$9.2 billion (64.3 billion RMB), according to the data from the mobile ticketing platform Maoyan. The 5.4% growth from 2018 is a sizable uptick for a market of China's scale (the second largest movie market in the world), but it is considered a slower growth from 2018, which had a 9% rise from the previous year. (Before that, China's movie market had enjoyed double-digit growth for most of the past decade.)

This trend may seem counterintuitive given today's explosive growth of online entertainment consumption. The Chinese market's rapid growth is in sharp contrast to the global box office, which sees movie attendance flat or declining in most markets. In other markets, the low-cost entertainment options on electronic devices have led to less movie-going for young people. For example, the 2019 box office of North America, the largest market of the world, slipped 4% year-on-year.

But it is truly an exhibition of the power of the internet and technology that keeps injecting new momentum into China's movie market. Just like other OMO cases like new retail, the movie industry is seeing smart devices, consumer Big Data analytics, ticket-booking mobile apps, and online movie streaming increasingly turning netizens into filmgoers at offline cinemas. Furthermore, the movies made domestically, not the Hollywood blockbusters, are the key driver for box office growth in China. Foreign films represented half of the top 10 movies in China in 2014, and they have made up a smaller mix ever since. The last time a Hollywood movie led China's box office was the 2015 *Furious 7.*

The secret? Big Data.

Until just a few years ago, China's movie studios controlled all business lines—from idea to script talent to production to theaters. Such practice was called "being a dragon from head to tail", where the studios take care everything and public audiences are passively on the receiving end. With the mobile internet providing Chinese consumers not only more content, but also more ways to find it, however, they have turned the table as choosers in consumption.

Today, consumers increasingly make their preferences heard by the directors and producers through social media and other channels. Like having dinner at a restaurant, the new generation of moviegoers would rather have their favorite dishes made to order than wait to see what the chef is cooking. Therefore, the studios proactively use Big Data analytics to find consumers' focus areas. The new studio model is: let us know what you like to watch (simultaneously we will try figure out your taste preferences by Big Data technology), and then we will produce it for you.

Therefore, for new films in China, their idea creation, production, and marketing campaigns are more and more shaped by Big Data, executed via social media, and distributed by the mobile internet (see Figure 3.9).

Finding the right story. Every day, millions of young digital-reading users refresh their mobile apps just to keep up with the latest daily updates of their favorite reads. For many people who do not have the time to read a book in hard copy, the novels on a mobile phone can be easily read whenever they have some spare time. Each serial installment typically has a few thousand words, so the reading can be done at any "fragmented time" in between everyday errands, such as waiting in line at supermarkets or for public transportation.

Based on their popularity among readers, many online novels have been adapted into traditional publications, games,

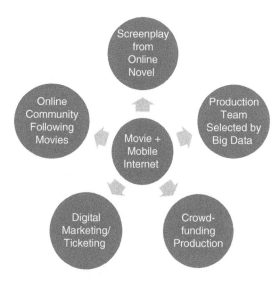

Figure 3.9   Data through the life cycle of a movie

videos, and blockbuster movies. The online literature's popularity rankings provide moviemakers with important information on potential moviegoers' constantly shifting preferences for storyline and genre; hence, more and more directors actively search the internet for potential hits (more detailed discussions of online novels are in Chapter 4 relating to mobile entertainment businesses).

Selecting the right production team. In a general platform like WeChat or specialized movie review sites like Douban, movie fans comment on the latest movies and make suggestions. For example, one of the hit movie series *Tiny Times* was based on a popular online novel. In addition to movie story choices, social networks also help shape the choice of the production team. Before Mr. Guo Jingming, the book author, became a first-time producer-director, he checked with his fan base for advice. The producing company also taps into Weibo (the Twitter of China) for suggestions on the director and stars of the films from more than 100 million Weibo followers. The feedback was somewhat dramatic because the public fans suggested Mr. Guo himself to be the director, and an important reason was "to preserve the original flavor of the novel".

Seeking continuous feedback during production. In a social general platform like WeChat or specialized movie review site like Douban, movie fans comment on the latest movies and make suggestions. Even after a decision on a movie production is made, directors may still use social networks to interact with fans because their feedback and comments not only help polish the original online novel for the movie screen, but also provide free marketing prior to film production.

Using digital financing channels. Movie crowdfunding has attracted many young audiences' interests and is quickly becoming an important funding channel for movie production companies. In the digital age, crowd-funding campaigns are marketed as an opportunity for ordinary audiences to get involved in the glamour of movie-making, since anyone who loves films or adores stars could become a minor investor and participate in movie production and promotion. For example, some investment products cover a group's movie projects. In order to gather information on audiences' taste preferences, each investor was asked to choose a maximum of two investment plans from the list of projects (more detailed discussion of crowd-funding is in Chapter 5 relating to the fintech revolution).

Leveraging digital technology for marketing. In the mobile era, the internet and smart devices are becoming the primary mediums between the movie market and Chinese audiences. Traditional marketing methods such as posters and trailers are not sufficient for large-scale film marketing and promotion. At the 2015 Shanghai International Film Festival, its organizing committee released its first ever report on the trend of "internet movies" in China, and its data showed that the internet became one of the most important marketing channels for Chinese movies, including 75.27% of Chinese film viewers decided to watch a movie because they read original network fiction, played related online games, or watched animations through the Internet.

Data on new media users makes it possible for precise marketing of films. For movies that target young audiences, productions spend less on advertising billboards across the country, and instead focus on the film's marketing activities on social networks. In smaller urban centers like tier 3 and lower cities, the internet may be the only cost-effective way to reach the potential audience. The mobile channels are especially important because for small-town youths, smartphones may be their only access to the internet. If young movie fans could be motivated to share movie information on the social network, the movie can potentially achieve far better coverage than by traditional offline advertisements.

Selling tickets on mobile apps. Mobile internet technology has also changed the way the moviegoers pick a cinema nearby and book movie tickets in advance. Movie ticketing was an area where the traditional movie theaters failed to provide convenient services in the past, due to a lack of investment in technology (movie theaters used to post their show timetable at the middle column or service page of local newspapers). Now the mobile ticketing service provides viewers with the most up-to-date movie schedules, location information of movie theaters, cheap tickets (thanks to the subsidies of the big internet companies), and seating reservation ahead of the show time. Collectively, the mobile apps have made an offer that the filmgoers cannot refuse, and the ticketing process collects more information of moviegoers for market analyses.

Bullet screen for feedback—even when the movie is on the screen. To provide more channels for young viewers to express their opinions, select movie theaters in China have experimented with the "bullet screen"—a controversial model that enables viewers to chat about the films via text messages on the movie screen as they watch them. **(See the "*Dan-mu* ['Bullet Screen'] and *Tu-cao* ['Spit and Criticize']" box.)** There is still debate on whether the bullet screen is simply a marketing trick or a major model revolution for the movie industry. For some audiences, the bullet screen is a distraction to movie watching;

for others, it is an addition of social interaction to enrich the overall movie-watching experience.

Most likely, the bullet screen, which could be viewed as the "digitalized shops" for the movie industry, is here to stay due to the young generation audiences' strong desire to express their views. To them, while the director defines the storylines on the movie screen, the bullet screen is a parallel screen of their own to provide feedback on the movies; to the production company, the bullet screen is another creative way to find out more about the audiences. In the future, it is likely that every cinema will set aside a special section dedicated to bullet screen viewers, and specialized bullet screen cinemas equipped with the cutting-edge technology may emerge like the IMAX theaters.

Dan-mu ("Bullet Screen") and *Tu-cao* ("Spit and Criticize")

Dan-mu ("bullet screen" in English) is a new model of movie watching. The term may originate from online warfare games, where intense shootings by cannons fill the PC or smartphone screen with bullets. Likewise, some theaters in China have launched a pop-up commenting section on screen, where viewers in the audience can comment on the film and have their text messages projected for everyone to see and reply. At any given time, those messages and comments scroll across the screen like bullets.

The "bullet screen" model was started at a video streaming site in Japan called *niconico*, and it has been adopted by major Chinese video sites such as Tudou and iQiyi. When watching a video on a mobile device, this technology allows viewers to send messages about what they like or dislike about the video, and all the people who are watching the video at the same time can see the message, respond to it, or post their own comments

In other words, the "bullet screen" feature of video streaming creates a virtual "watching together" experience among people, even though everyone is watching individually. The intense real-time social interaction among the viewers provides an entirely different experience from the traditionally web setting where viewers post their comments on the web page (not the screen) in sequence (not real-time). While comments can be so overwhelming that the content screen itself could get clouded, "bullet screen" proponents suggest the highlight of the experience is not the video itself, but the *tu-cao* (Chinese for "spit and criticize in a joking way") in a virtual social context.

Now the movie marketing people have bravely applied the "bullet screen" model to a real cinema setting, and recent experiments included major productions such as *Tiny Times*. Probably due to the self-selection of the audience attending those cinemas, the reactions to those trials were generally supportive.

For example, many viewers seemed to be excited when seeing their comments shown on the big screen. Some praised the model for combining watching movies and making friends at the same time. Some thought it was an especially good idea for boring or silly movies, so that instead of passively viewing the content, they could still enjoy the social experience of *tu-cao* the film jointly. Some even suggested that the movie plots were occasionally hard to understand and the messages on the bullet screen provided insight for them when they were struggling to understand.

The "mobile internet+" and Big Data trends in China's film industry well explain why Hollywood blockbusters, characterized by big-budget productions and visual spectacles, have lost market share during the tremendous growth of China's film market in recent years. Local films are catching up to Hollywood on many indicators like story, special effects, and genre, while Hollywood films are struggling with self-repetition and a lack of originality. Over the past three years, foreign films have accounted for a shrinking slice of the box office (see Figure 3.10), with Chinese films performing proportionally better. In 2019,

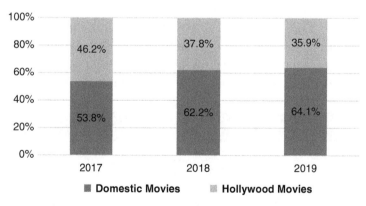

Figure 3.10   The declining China market share of Hollywood movies
Data Source: Maoyan.

Chinese films accounted for 64.1% of the total box office—an increase on the 62.2% in 2018 and 53.8% in 2017. (In the past, domestic movies only claimed a small market share due to the dominance of blockbuster imports from Hollywood.)

In particular, just as in the case of Pinduoduo, China's domestic films have adapted to the rapid increase of the "small-town youths" at cinemas. This important group represent a void in the traditional domestic movie market. On the one side of the spectrum of available movies were those from older authors and directors in China, to which the young people could not relate or did not even try to understand. On the other end of the spectrum were the foreign movies that appeared too remote and random for the "small-town youths." Most members of this young audience enjoy stories from authors of their own age, admiring fashionable, popular, and successful heroes who are also homegrown.

For example, in the case of *Tiny Times*, the luxurious lifestyle of high society girls in the movie, from fashion brands to exclusive clubs, clearly resonates with the young audiences' imagination of success. In 2019, the movie that topped China's box office was *Ne Zha*, a 3D fantasy animation, which has become China's most successful animated movie ever (US$703 million). The film is based on the mythological character of Ne Zha, a deity well known in China for being a rebellious teen hero. The protagonist, as a demon reincarnation, is destined by prophecy to bring destruction to the world, but fights against his destiny—a theme echoing with the large part of Chinese population who are fighting social stereotypes.

Xiao-xian-rou ("Little Fresh Meat") Actors

The term *xiao-xian-rou* literally means "little fresh meat", which refers to young men who are youthful, innocent, and beautiful in both face and body. The term started from the media industry, and it is now also being used generally for similarly attractive figures in other fields, for example, young swimming athletes and professional dancers.

There are several important characteristics of *xiao-xian-rou* actors. First, they are young and energetic ("little"); second, they are relatively inexperienced in love affairs and have no sex scandal news ("fresh"); and third, they must have strong muscles while they are enviably good-looking ("meat"). Further on "meat", they should also have impeccable skin—perhaps even better than that of girls.

Clearly, the term has a lecherous and commercial connotation. It expressively describes the physical body of youth that are desirable by older members of the opposite sex. But if *diao-si* (meaning losers), a term based on the male genitalia, could become a widely accepted term that occasionally even shows up in news headlines, who should be bothered by the *xiao-xian-rou* reference?

It seems that no one is unhappy when being called as such because the term is viewed as praise to the person's youth, energy, and beauty. In fact, some established actors have expressed the hope to remain as young as the new generation of *xiao-xian-rou* actors.

The ultimate question, however, is on the sustainability of this exceptional growth. The movie audience is maturing. Many are less willing than before to buy tickets to movies with only visual spectacles and young *xiao-xian-rou* actors, which requires ever more Big Data analysis of the changing moviegoers. (**See the "*Xiao-xian-rou* ['Little Fresh Meat' Actors]" box.**) Guided by the Big Data of their users, all the e-commerce giants have set up their own movie units to produce movies, especially those narrating stories of interest to this young generation.

Still, will online entertainment eventually take away audiences from the big screen as in developed markets? So far there is no data pointing in that direction, as the Chinese movie fans seem to want both the cinema and the mobile-device experience. As big movie screens and the mobile internet continue expanding the reach into small cities and rural areas, more "small-town youths" become new moviegoers, and the market likely will see record-setting box office numbers continue in the coming years, albeit at a slower pace than before, as China has already become the country with the most film screens in the world at 70 000 by 2020 (North America is estimated to have around 46 000).

Because of the coronavirus (COVID-19) pandemic, China's movie market came to a screeching halt in early 2020, only to recover slowly in the second half. Consequently, it is not the year for China to continue its growth and surpass the United States to become the world's largest box office market. However, the movie production has not stopped. New movies are being launched on video streaming sites, which accelerates the fusion of big (cinema) screens and small (smart device) screens. Chapter 4 covers the convergence of various mobile entertainments on multiple screens, and the "bullet screen" feature is in fashion.

Post-pandemic Consumers Come of Age

In this chapter, the case studies of *Hema* "new retail", Pinduoduo C2M social e-commerce, and "Big Data" movies are all successful examples of consumer data-driven OMO models. The challenge to the corporate world, however, is the ever-changing consumer trends. Since early 2020, China has been on the frontlines both of coronavirus-driven economic downturn, and of the societal changes the pandemic has precipitated.

The impact of the coronavirus crisis on Chinese consumers has been profound, especially for the affluent younger generation that has never experienced a domestic economic downturn. Before the epidemic, China's young consumers were more willing to spend than to save. As they come to terms with the prospect of recession for the first time in their working lives, they are forced to think harder about spending, borrowing, and trade-offs in purchasing behavior. Some may plan to save more (or borrow less), some may spend more money on eco-friendly products, and a lot are saying that they want to eat more healthily after the crisis. Meanwhile, as expected, everyone is buying everything online.

Therefore, more businesses most likely will go digital and consumer data-driven after the COVID-19 crisis. The good news is that more advanced digital technologies are becoming available to corporations to better profile their consumers and to seamlessly cover the 4Cs—Context, Community, Content, and Connection as will be illustrated in more case studies in the rest of this book.

The AI-powered Internet Celebrities and Fans Economy

- Bilibili: Bullet Chats, Membership Questionnaires, and Gen Z Uploaders
- Tencent's China Literature: The Gods of Internet Novels
- Beyond Twitter—The (Brief) Revival of Weibo's UGC Platform
- TikTok: Goofy Videos Dominating Social Media
- Kuaishou, IQiyi, and More: Into the Multiple Screen Competition

Bilibili: Bullet Chats, Membership Questionnaires, and Gen Z Uploaders

How much would it cost a company to host the coolest New Year's party? (Hint: do not think like a Hollywood agency or Wall Street.) The best answer: a US$1 billion increase in the company's market valuation.

Amid all the 2020 New Year's Eve countdown shows, Bilibili's online New Year Concert went viral, notwithstanding being the only one among eight New Year's shows in China that wasn't broadcast on TV. Bilibili, started in 2009, is China's biggest anime, comics, and games (ACG) content community as well as one of the country's biggest video-sharing sites. While the New Year's festivities have traditionally been monopolized by China's large state networks, the Bilibili concert won the hearts of Gen Z and millennial audiences with innovative formats, anime songs, and plenty of meme references.

The pre-recorded stream kicked off with an O-DOG World of Warcraft themed dance, in honor of faithful fans of the 15-year classic game. If viewed from the perspective of traditional gala shows, it would be a ridiculous opening. The dance was followed by the 2019 League of Legends World Championship theme song "Phoenix". Instead of mainstream singers, the show featured online celebrities from the "music" category of its site (so-called "top unloaders"), such as Kris Wu, whose performance was based on his widely ridiculed freestyle rap about noodles. During the runup to the New Year, the lineup also included Harry Potter, *Ne Zha* from China (the best box office movie of 2019; see Chapter 3), and the hit animation *Spirited Away* from Japan.

On New Year's Eve, the Bilibili show reached 82 million simultaneous viewers at its peak, and a further 40 million playbacks had been recorded by the following day. Listed on the Nasdaq stock exchange in New York, Bilibili's stock powered ahead after New Year's Day. On January 2, 2020, the stock jumped 12.51%, followed by another 5.39% on January 3. The company's market capitalization increased for more than US$1 billion on these two trading days only. In April 2020, Sony Corporation invested US$400 million for just under 5% of Bilibili's equity ownership. In addition, Bilibili and Sony entered into a business collaboration agreement to pursue opportunities in the Chinese market, including anime and mobile games. The momentum carried on and the company's valuation exceeded US$10 billion on Nasdaq in the first half of 2020.

The Bilibili show and its platform represent a powerful trend: China is becoming the world's largest online entertainment market and the young generation is at its center. According to the CNNIC data, by March 2020, more than 40% of the Chinese internet users are between ages 10–29

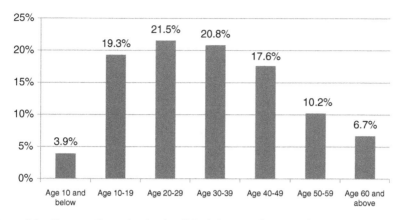

Figure 4.1   Young netizens dominating China's internet Demography
Data Source: CNNIC Report, March 2020.

(see Figure 4.1). This increasingly open-minded, aesthetically refined, playful, and fluid group is evolving from the cultural niche to mainstream in China. They are the driving force and trend-setters of entertainment consumption. As such, "[O]ur mission is to enrich the everyday life of young generations in China", said the company, which builds its platform as a destination to discover cultural trends and phenomena for young generations in China.

According to QuestMobile's report, the majority of Bilibili user base (over 80%) is Generation Z, individuals born from 1990 to 2009 in China. They typically receive a quality education and are technology savvy, with a strong demand for culture products and avenues for self-expression and social interaction. Interestingly, just like in the movie industry (see Chapter 3), the most rapid growth of user population is also from the "small-town youths". Before the widespread use of smartphones and national broadband infrastructure, the young generation from China's vast lower-tier cities and rural places was an uncovered group. Now these young people are experiencing a consumption upgrading trend where online entertainment becomes more and more of an urgent need.

The most significant change that digital technologies have brought to the fans' economy is that the fans are, for the first time, at the center of online entertainments. At new media platforms, thanks to interactive features, and more recently, DIY video-making tools, fans are no longer passive consumers of entertainment content. As a result, new social media platforms are now pivoting from "elite"-centered community toward the untapped grassroots crowds derived from smaller cities, younger ages, lower social statuses, and less glamorous–looking groups.

Whereas many fans self-jokingly call themselves *diaosi*, they nevertheless aspire to join the production of viral entertainment contents. In fact, showing off (or *shai*) their lifestyle online is part of the social fabric for many young Chinese people. **(See "*Shai* ['Showing Off'] and *Daka* ('Punching the Card']" box**.) In order to engage them effectively, the media platforms in China no only provide the fans a platform to find and consume contents, but also equip them with digital technologies to create and contribute contents of professional quality, which is a common theme in all the case studies in this chapter.

Shai ("Showing Off") and *Daka* ("Punching the Card")

Shai in Chinese means "to put something under the sunshine". In social media, describes members of the young generation "showing off" their lifestyle, such as a luxury brand they just bought or a fancy restaurant that they're patronizing.

For example, foreign cosmetics brands have been a popular category for Chinese consumers to *shai* when they enjoy "trading up" into luxury brands. For brands, *shai* means the young netizens are not passive consumers; instead, they actively find a voice in social media to express their appreciation of the brands (or they may similarly express dissatisfaction with brands on their social network). Nowadays, more often, this young generation *shai* their unique experiences, thanks to the development of video technologies.

Instead of having fun, some people now flock to remote fruit farms or photogenic spots simply to upload a picture or video to media platforms such as Douyin (Chinese version of TikTok) and Kuaishou – to impress their social network. This leads to a new subculture called *daka*, meaning "punching the card". The word is used to refer to the practice of registering one's presence at a popular location. Neither the beauty of the attraction or a well-crafted video or beautiful photograph is important. The aim is to show that the person has been to the same cool places as his or her peers.

Often the young people *daka* together as a lifestyle, forming so-called *daka zu* ("*daka* tribes"). They can be found roaming cities, punching the card at as many hot locations as possible within a single day. A big business has consequently emerged to assist the *daka* tribes. Guides can be found online to identify the most efficient ways to achieve the most card punching. Travel agencies offer "*daka* tours". Also, video streaming apps like Douyin empower their users to create "*daka* videos": super-speed slideshows of themselves at hot locations.

The *daka* phenomenon may have a practical reason: the young generation is working too hard. Many of them have the famous "9–9–6" work schedule – 9 am to 9 pm, six days a week (see more details relating to China's innovation ecosystem in Chapter 7). They have little vacation time, so they must make the most of their limited leisure time.

Bilibili, for example, captures the hearts and minds of its young users with carefully designed interactive features. The platform's most celebrated feature is "bullet chatting", which allows users to post "bullet" messages in response to live videos. Every viewer sees time-synced comments, thoughts, and feelings of other people viewing the same video overlaid directly on top of the video as it plays. This signature feature fosters a highly interactive viewing experience and allows its users to benefit from the strong emotional bonds with other users who share similar aspiration and interests.

As such, young fans go to Bilibili for the community as much as the videos themselves, whether anime or music, comics or games. During pivotal moments in a video, reactions wash over the video in a dense tidal wave, often covering up the screen across a whole face. (For those young audiences, their approach to video watching can be best summed up by the title of a 2010 Chinese action comedy: *Let the Bullets Fly*.) To the comfort of the less trailblazer users, the company utilizes an artificial intelligence-based screening system to conduct semantic analysis on bullet chats to screen out ones that are inappropriate.

Since Bilibili began its annual bullet chat ranking in 2017, the top 10 results have provided a unique insight into the cultural trends of China's Generation Z youth. The inaugural winner of 2017 was 囍 (*xi*) ("double happiness" in Chinese), and the year 2018's was "real". The "2019 Bilibili's Bullet Chat of the Year" was the term *AWSL*, which came from "Ah, Wo Si Le"! in China—meaning "I'm dying".

AWSL is similar to "This is killing me" in English, often used on social media to describe something extremely cute, exciting, or funny. It is used to express someone's unique amusement when watching a Bilibili video. "Snippy and spontaneous, AWSL reflects contemporary youth culture, a great medium for online interaction", analyzed Gao Hanning, a post-doctorate at the Chinese Academy of Social Sciences Institute of Literature, in related media reports. (See Table 4.1 for the complete list of Bilibili Top 10 Bullet Chats in 2019. For readers who are not young enough, the Chinese characters, English transaction, and emotions expressed are all included for cross references.)

Table 4.1   Bilibili top 10 bullet chats in 2019

Ranking	Bullet Chat	Translation	Emotions Expressed
1	"AWSL"	"I'm dying" or "This is killing me"	Something especially cute, exciting or fun
2	"泪目"	"Teary-eyed"	A touching or sentimental moment
3	"名场面"	"Iconic"	A classic or well-known scene
4	"妙啊"	"Marvelous"	Something that is original, ingenious, or fantastic
5	"逮虾户"	"Déjà vu"	Mostly used when viewers react to scenes of cars racing or drifting
6	"我可以"	"Yes, I can"	A phrase used when one is strongly interested or excited about someone or something
7	"欢迎回家"	"Welcome home"	A phrase meaning "you are not alone"
8	"注入灵魂"	"Soul-injecting"	Adding a finishing touch or something unique in the process of creation to elevate the quality or significance of the work
9	"正片开始"	"Official start"	The moment when the plot of a video or game begins to hit the peak of its story arc
10	"标准结局"	"Standard ending"	A disapproving reaction to the expected ending of a video or game

This bullet chat feature is similar to the "bullet screens" in some Chinese movie cinemas (see Chapter 3 for discussion relating to the "Big Data" movies). However, not everyone on the website can shoot—you must pass the entrance exam first. Any user who visits the platform can watch or search content, and then he or she must register to activate the basic interactive features on the platform, such as liking videos and following content creators. Additional interactive features, such as bullet chatting and commenting, become available to registered users once they become the "official members" by passing a multiple-choice exam consisting of 100 questions.

The membership exam includes questions on community etiquette regarding uploading videos and sending bullet chats, and a set of questions on a range of topics with which registered users are (and should be) familiar, such as anime, music, games, and technology. Registered users need to answer a total of 60 questions correctly to pass the membership exam. As of the end of 2019, Bilibili had 67.9 million official members who passed its membership exam. This rite of passage encourages a sense of exclusivity among those who make the grade which, in turn, creates a highly sticky community.

Furthermore, Bilibili has developed itself from a source for ACG (anime, comics, and games) to a thriving community of homebrew video stars. As is seen in the different case studies across multiple content formats, the ever-growing supply of professional (quality) user generated content, or PUGC, is a common theme in all online entertainment platforms. PUGC combines the content breadth offered by user-generated content (UGC) and the quality offered by professional-generated content (PGC). Its emergence highlights that the users are content creators themselves, which makes them the unequivocal center of those platforms (vis-à-vis entertainment companies that supply big-budget productions).

Professional user generated videos, or PUG videos, have recently emerged as the most popular category of content. On the Bilibili platforms (and all other entertainment platforms), videos are now created by a wide range of participants, from amateurs to professional users who have a certain level of production and editing capabilities, and to professionals from production studios or workshops, but the lines that separate their two contents are increasingly blurring. In the past, UGC was typically viewed as low in quality; however, with the development of affordable and easy-to-use hardware, such as digital camcorders and mobile devices with high-resolution video cameras as well as advances in software technology such as desktop editing software, the barrier for producing quality video content is rapidly vanishing.

Bilibili has developed a variety of uploading tools to enable users to efficiently upload multimedia content, including videos of indifferent length, pictures, blogs, and other forms of content. Some of the uploading tools also contain editing features that can help users add a variety of visual and audio effects to the content. As such, Bilibili has its own community of content creators called "uploaders". Their PUG videos are well received by Bilibili users due to their originality as well as their sharing and interactive characteristics. In 2019, PUG video views accounted for approximately 90% of Bilibili's total video views, which explains why, for the 2020 New Year Eve's gala, the singers and performers were mainly selected from the "top uploaders" (the most popular content creators on the platform).

In addition to tech support, Bilibili further incentivizes "uploaders" to create content by facilitating e-commerce opportunities for them. Whatever field a brand is in, there is probably a route to engaging China's Generation Z consumers by leveraging content on Bilibili. In December 2018, the company and Alibaba's **Taobao Marketplace** entered into a business collaboration agreement in content-driven e-commerce. Under the agreement, the two companies collaborate to develop a dynamic ecosystem that will better connect content creators, merchandise, and users on both platforms.

As part of the partnership, the online marketplace will introduce content creators from Bilibili to the international brands that are working with Taobao. Bilibili has popular uploaders across fashion, beauty, travel, health and fitness, child and parent, home and design, automotive, and other categories who can help international brands running marketing campaigns. On the 2020 New Year's Eve, for example, Taobao's flash-sales platform **Juhuasuan** was the title sponsor of the Bilibili gala. To target Bilibili's users, Juhuasuan released special products with ACG themes. (Also in 2018, Tencent invested US$300 million in the company following its Nasdaq listing in March.)

In summary, the three forces of young users, mobile payment, and e-commerce have jointly ignited the fans economy in China (see Figure 4.2), creating numerous dynamic platforms to be discussed in this chapter. The fans today take an active and direct role in content creation. Becoming an online celebrity is a motivation for many users to create viral-worthy unique content of their own, and the new digital technologies have empowered them to achieve that much easier than before. Thanks to advanced technologies readily available to ordinary users, the line between UGC (User Generated Content) and PGC (Professional Generated Content) is also disappearing as PUGC is no longer the synonym of low-quality and low-taste content.

Meanwhile, the entertainment fan economy can be characterized as highly competitive and rapidly changing due to the fast-growing market demand. With the growth rate of the overall size of the internet community slowing down, the industry is witnessing rising competition for user traffic and attention time, as exemplified by the rise of short-form video platforms,

Figure 4.2   The three key components of a fan economy

such as Douyin and Kuaishou. As is discussed in the following sections, every player strives to offer a full spectrum of entertainment content, including PUG videos, licensed videos, live broadcasting, short video clips, pictures, blogs, games, and any new media format one can imagine.

For example, during the 2020 pandemic, Bilibili stretched into longer format films and documentaries. It recently partnered with Discovery and China Intercontinental Communication Center to produce the *COVID-19: Battling the Devil* documentary about China's fight against the coronavirus. Furthermore, although the Generation Z provides the most growth, all the multimedia content platforms are also fighting for users of all age groups. Bilibili's 2020 New Year's Eve Gala, for instance, not only exemplified its deep understanding of young people's interests, but also was designed to impress generations born in the 1970s and 1980s (some quickly joined the discussion in social network and eventually took the entrance exam).

The following sections examine Tencent's China Literature (the largest online literature platform), Weibo (China's equivalent of Twitter), Baidu's iQiyi (China's equivalent of Netflix), ByteDance's TikTok (currently the most valuable unicorn in the media sector), and more platforms that share the same aspiration to be the one-stop entertainment platform for the young generation—and everyone else. Video streaming, especially short videos, is at the center of the mobile platforms and also the focus of this chapter. However, internet literature is at the upstream of the flow of IP (intellectual property) rights in the industry as it is the base for a variety of derivative contents, including publishing, gaming, animation, and filming as well as videos. Therefore, the online literature business is introduced first.

Tencent's China Literature: The Gods of the Internet Novels

Online novel reading is popular in China, partly because it is one of the cheapest forms of original entertainment content. The CNNIC data showed that as early as 2014, most

Chinese readers had already migrated to mobile devices, with approximately 85% of readers using mobile smartphones to read literature works, far ahead of the usage of other reading mediums like printed books or PC computers. By the end of March 2020, there were more than 450 million online literature readers, representing approximately half of the total internet population.

Shanda Literature used to own the most popular websites and nearly 50% of the market. Launched in 2008, Shanda was a pioneer in enabling Chinese writers to freely publish online and monetize their work through micro e-payments from readers as well as rights and licensing deals. Despite its leadership in content, Shanda Literature missed out on the development of the mobile reading market, and in early 2015, it was acquired by Tencent Literature, whose parent Tencent has mobile entertainment in its corporate DNA. After the merger, the new entity, Yue-wen Group (meaning "Reading Literature" and its English name is **China Literature Limited**), became a superpower in the online literature field with approximately 70% of the market share. In November 2017, China Literature was listed in Hong Kong, and the stock price soared as much as 90% on day one of trading in what's been dubbed the city's most-profitable IPO in nearly a decade.

Today, the company remains China's largest online publishing and e-book website by the scale of writers, readers, and literary content offerings. The company owns nine major branded products, among which is the flagship product QQ Reading, a unified mobile content aggregation and distribution platform. As of December 31, 2019, the company had 8.1 million writers and 12.2 million online literary works, covering over 200 genres. Thanks to the user traffic from the ecosystem of Tencent—the largest shareholder that held a 57% stake in the company at the end of 2019—the China Literature platform has more than 200 million monthly active users. According to Baidu's February 2020 search rankings,

25 out of the top 30 online literary works were created on the platform.

In spite of its humble format, the internet novel business is a perfect example of the 4Cs model from Chapter 3, seamlessly integrating users around the 4Cs in the digital economy: Context, Community, Content, and Connection. To start, the unique feature of Chinese online literature is that most works are serialized novels that authors write and post in installments (a new form of "**Content**"). This online serial format proves a perfect fit for the mobile internet age. Every day, millions of young digital-reading users refresh their mobile apps, just to keep up with the latest daily updates of their favorite reads.

For many people who do not have the time to read a book in hard copy, the novels on a mobile phone (a new "**Connection**") can be easily read whenever they have some spare time. Each serial installment typically has a few thousand words, so the reading can be done at any "fragmented time" in between everyday errands, such as waiting in line at supermarkets or for public transportation (various "**Contexts**"). In chat rooms and on social networks, millions of fans actively comment on the story plots and discuss them among themselves (a new

Figure 4.3   Online novel business representing the 4Cs model

"**Community**"), and the novels' popularity is digitalized and quantifiable.

When authors start to build up large readerships, the online portals can offer them contracts and move their works off the free domain. The installment format helps the literature websites to implement a pay-for-content mechanism, and the development of mobile payment systems (another new "**Connection**") makes it convenient for readers to make small and repeating payments for their serial reading. The sites arrange for the authors to write a story in installments (typically with a cap for a total number of characters for each post), and readers then pay a tiny fee equivalent to a fraction of US$0.01 to read each update, far cheaper than paying for hard copy versions from a bookstore.

Interestingly, the wish-fulfilment themes are a major reason that the online novels are so enormously popular among the young generation. A big chunk of the readers is in China's smaller cities, calling themselves *diao-si* ("losers") readers for not owning an apartment nor having a girlfriend. The heroes are handsome, powerful, and successful, and the sensational plots of the novels provide for the readers a context to dream of being the heroes themselves. For that, the internet authors have created many imaginary worlds for the enjoyment of their readers. (**See the *Diao-si* ['Loser'] Readers and *Pu-jie* Author ['Drop-Dead-on-Street'] Authors" box**.)

The same dynamics also played out during the COVID-19 pandemic in 2020. "Online novel readers can find comfort in a literary world that takes them away from reality. Stay-at-home arrangements have given them more opportunities to read; reading is a kind of emotional vaccine that help them fight the virus," a professional fiction writer who is a contracted writer with ByteDance's Tomato Novel, said to the news media during the coronavirus crisis. (See more details about the Tomato Novel in the section relating to TikTok, its sister product.)

Diao-si ("Loser") Readers and *Pu-jie* ("Drop-Dead-on-Street") Authors

The term *diao-si* has appeared earlier in this book, and in general, it refers to unmarried young people who live on the margin in an otherwise booming economy. In the context of e-commerce, the term is related to little spending power and cheap goods. For example, the users of non-premium-brand smartphones tend to call themselves *diaosi*.

In the context of online literature readers, the small installment payments for escapist content certainly fit this description. But when the online readers dub themselves as *diaosi*, they are also highlighting their dream toward a better life. In fact, this aspiration is evident in fans in all entertainment fields, whether games, animations, videos, or movies. They admire fashionable, popular, and successful heroes that are also home-grown, and the successful local dramas in recent years centered on individualistic pursuits for happiness and success.

For example, in the case of the *Tiny Times* online novel and then movie series (see more details in Chapter 3), the protagonist's path was every Chinese college student's dream life: she began the story as an ordinary university student, and then she had to face all these things like job interviews and then intimidating bosses in the workplace. But she managed to pull through and her life became better and better—she found a glamorous career, great friends, and a handsome boyfriend.

It turns out that some online authors are in no better living condition than the *diao-si* readers. The millions of web writers in China are divided into multiple categories by their income and number of fans. The lowest level is called *pu-jie* authors, which can be translated into "drop-dead on street" authors. Their fans are limited, their work is seldom recommended online, and their annual income is in hundreds of dollars. China Literature reported in early 2020 that among 8 million registered authors on its platform, the vast majority earn a mere 1000–2000 RMB (approximately US$150–$300) each month for writing over 4000 words a day for 30 days straight.

In contrast, the successful authors in real life are no *diao-si* at all. Some internet authors are celebrities both online and offline as their popular novels have made their way into films and TV programs. Just like the stars on Google's YouTube, they leverage their online celebrity in various parts of the media market like publishing, filming, and live events. The highest-paid online authors are called by the *diao-si* readers "the Supreme Gods". However, according to the China Literature data, only about 428 authors among the 8 million are known as "Platinum Authors" (equivalent to the "Major Gods" in Table 4.1), and even much fewer reach the status of "Phenomenal Authors" (equivalent to the "Supreme Gods").

Furthermore, the *diao-si* readers feel even more empowered when they can change a writer's story direction by posting their feedback and recommendations. Most internet literature websites include a reward function, where readers can award the web writer money or credits out of discretion. If a reader is especially satisfied by the story plot's change based on his or her suggestion, he or she could make a one-time big tip (a "gift") to the author. Thanks to technological innovations, the China Literature platform has recently rolled out new features to enable readers to participate in the creation of stories they like:

- **More instant comments.** A function launched in 2018 is "paragraph commenting" (a comment function for as small a unit as a paragraph), which enhances user engagement with the writer by encouraging more instant comments from readers.
- **Your own voice (literally).** A new function rolled out in late 2019 allows users to submit their own audio recordings of select sentences in a novel and listen to and comment on others. This popular function brings users' voices to the original text and creates more layers of context for engagement, which brings the community closer together.
- **Improved recommendation efficiency.** In addition to building a social graph within the user community, the platform also uses deep learning and natural language processing technologies to map out the interests of individual users and features of literary works so that it can recommend the most relevant content to the users.

Despite readers' increasing participation, their feedback may not always be able to change the stories because the authors had planned the novels before they wrote them. By the time the fans read the latest installment, the author may already be working on future ones. But that would not necessarily put an end to *diao-si* readers' dream: the readers can turn themselves into authors of their own stories. That is the biggest difference between the online novel and traditional book publishing. To publish an online novel is not like writing a book, where the author must complete the plot and write all the chapters before being published. Instead, a user only needs a few thousand words to start off a book and to show readers what is being written.

As such, it is quite easy for fans to become authors themselves, making the online novel sites dynamic and sustainable. Because the internet has leveled the playing field for aspiring young writers, even unknown authors could show up with their popular hits. For example, before the internet age, it was really hard for young screenwriters in China to emerge because a handful of producers working with a small number of screenwriters control almost the whole movie market; these days, however, many online novels from first-time authors have been adapted into feature movies.

The millions of online writers in China are categorized by their readers into five levels based on their income and number of fans, according to a recent survey by Beijing-based newspaper *Jinghua Times*. Except for the lowest rank *pu-jie* authors, which could be translated into "drop-dead-on-the-street authors", all the professional writers in higher categories are all called "gods." But they have different titles, like the deities of the Greek pantheon (see Table 4.1; on the China Literature platform, the "Major Gods" and "Supreme Gods" are referred to as "Platinum Authors" and "Phenomenal Authors", respectively. The "Middle Gods" and "Small Gods", however, form the core of the online novel ecosystem).

Similar to Bilibili in the previous section, the China Literature platform is increasingly driven by the young Generation Z. In 2019, much of the innovation in terms of content is

Table 4.1   The five ranks of web writers in China

Rank	Number of Fans	Annual Income (RMB/US$)	Number of authors
"Drop-dead-on-the-street" authors	Few fans, only recommended occasionally	< 1000 RMB ($150)	Numerous
Xiao-shen (Small God)	> 100 000	> 100 000 RMB ($15 000)	Large number— foundation of sites
Zhong-shen (Middle God)	> 500 000	> 500 000 RMB ($80,000)	Several hundred
Da-shen (Major God)	> 1 million	> 1 million RMB ($150,000)	Several hundred also
Zhi-gao-shen (Supreme God)	Multiple millions	Multiple millions RMB (>1 million US$)	20–30

Data Source: 2015 survey by Beijing-based newspaper *Jinghua Times*.

contributed by Generation Z writers who have the passion and ability to engage with younger demographics in a distinctively different manner than previous generations. They account for 25% of the Platinum and Phenomenal writers, and push the boundaries of the online literature world. For example, the author of *Lord of the Mysteries* managed to creatively combine elements of drama and video games to tell his story in an unconventional way.

In a broader context, because of the huge size of online novel fans and their high overlaps with those of other media contents, the novel fans' influence goes beyond the reading market. Typically, the devoted fans of literary works are not only readers; they are also inclined to be loyal audiences of dramas and movies adapted from the literature, or players of games and fans of animation. As such, internet literature is the base of IP (intellectual property) rights for a much larger fan economy than it first looks (see Figure 4.4). Therefore, it is a key piece for entertainment platforms, not only providing upstream intellectual property (IP) rights for inter-linked entertainment products, but also allowing the platforms to monitor the trend of user tastes.

Figure 4.4   The Internet novel as the "upstream" IP

China Literature, for example, took a major step toward bringing its IPs to life through a drama series with the acquisition of **New Classics Media** (NCM) in October 2018. NCM was considered to be the critical missing piece to amplify the value of its IPs, and after the acquisition, China Literature is positioned to be a leading drama series and film production studio. Based on the online literature libraries, NCM has developed top-tier content with the release of *Memories of Peking, Awakening of Insects, Joy of Life,* and *The Best Partner* throughout 2019.

All these drama series ranked top in terms of viewership during their respective broadcast time slots. Most notably, *Joy of Life,* which was adapted from one of the platform's most popular novels, ranked first among all TV and web series on Baidu and ByteDance Toutiao's 2019 search indices. Interestingly, the success of the drama series has also rekindled interest in the novel, which once again topped the rankings on the best-seller list since its original launch over 10 years ago and attracted 3.5 million recommendations and over 600 000 rewards. Building on this phenomenal success, development for *Joy of Life* Season 2 is already underway.

Figure 4.5   Ecosystem—Prolongs the life cycle of an online novel

Furthermore, China Literature recently formed a strategic partnership with **Tencent Music**, a leading online music entertainment platform in China. The strategic partnership allows Tencent Music to produce audiobooks for online literary contents. These cross-format content developments, whether from novel to drama, audio, or movies; whether in partnership with Tencent affiliates (Tencent has its own movie making unit, Tencent Pictures) or third-party players, can bring organic growth to Tencent's overall media ecosystem. Essentially, they are creating a virtuous cycle for premium IP incubation (see Figure 4.5).

This strategic direction shift into a focus on "cross-format" IP is most evident since 2020. In a power move widely regarded as a Tencent takeover, China Literature's senior management team, consisting mostly of founding members, resigned from their posts in April 2020 and were replaced by veteran executives from its controlling shareholder Tencent. The immediate impact was that China Literature is now moving toward a free-to-read model completely.

As the total online reader population in China has plateaued since 2018 and online readers' growth slowed, China Literature in 2019 started to allow price-sensitive users to read selected literary works for free while the platform monetizes through advertising. While the platform continues to support writers who monetize their efforts via the pay-to-read model, its revenues from IP operations and others has exceeded the revenue from the paid reading business, partly due to the integration with NCM businesses, as the platform diversified further into licensing of IP for adaptation into films, TV and web series, animations, and games.

For a platform with the most contracted drama writers, it was simply natural to see a drama developed between China Literature and the authors themselves. A strike, named "No Updating Day", was organized on May 5, and received the support of thousands of authors on the platforms. Hundreds of thousands of posts were sent with the hashtag, generating more than 100 million views in total on Weibo (China's main social network to be discussed in the following section). According to the authors, had the company removed the paywall on all content across its sites, their income, which primarily came from subscription fees and rewards from their readers, would have been greatly reduced.

In the end, the controversy is about how to divide a bigger pie. If the cross-format ecosystem would reap more revenues from more sources (advertising, adaptation, and licensing), prolong the life cycle of the novels, and save the authors' concerns of copyright infringements (some readers always find a way to read the contents behind a paywall for free), that could be a better deal for the authors. As is seen in this chapter, the free-to-read model is set to happen across all platforms. When the stakeholders reach a new and more profitable balance, the dreams of authors and readers can go on happily thereafter.

Beyond Twitter—The (Brief) Revival of Weibo's UGC Platform

Sina Weibo was launched in 2009 as a micro blogging service (which literally translates into its name *weibo* in Chinese) and mostly known as "China's Twitter". Like Twitter, Weibo also worked as a follower/followee network; it used similar opinion leaders in China such as media stars and business leaders to attract the early user base, and Weibo originally had a 140-character limit for posts as well (which was removed earlier in 2016 when it reformed its business model).

Since 2016, however, Weibo has transformed itself into a comprehensive media platform, combining all the contents and features of social media channels like Twitter (text), YouTube (videos), and Instagram (pictures). It combines the means of public self-expression in real time with a powerful platform for social interaction, content aggregation and content distribution. Ever since, the company had the fastest growth in its business in the years 2016–2017, and its stock price (on Nasdaq) jumped more than five times (see Figure 4.6). During the same period, Weibo's market valuation exceeded Twitter. The following Weibo case provides a useful background before the discussion of new media platforms in the rest of this chapter.

Weibo's growth came as a big surprise to many. For one thing, the video streaming industry trend was that video content was shifting from user-generated content (UGC) to professional-generated content (PGC) as viewers increasingly

Weibo Stock Price Movement

Figure 4.6   Weibo's stock price run (2016–2019)
Data Source: Capital IQ.

demand high-quality content. For another, because Tencent's social messaging app WeChat gained wide popularity years ago, many people switched from Weibo to Wechat as the latter became the more popular platform for the public's need for a social network, mobile messaging, and interactive private conversations. As a result, Weibo's growth tapered off in the years after WeChat's appearance, and many predicted that Weibo's time was over.

Then where did Weibo's huge revival come from? The answer is very simple: users demanded more diverse content consumption and more active participation in content creation, and Weibo correctly responded. To satisfy users' growing needs, the company made a critical adjustment to its positioning—Weibo set itself as a social media company, not a microblogging site. To that end, Weibo's new approach was on building simple and useful tools to enable its users to discover, create, and distribute content, publicly express themselves, and interact with others on the platform in real time (see Table 4.2).

Table 4.2   Weibo's critical adjustment to its positioning and offering

	Before (Twitter of China)	Now (Multimedia Platform)
Interface	PC	Mobile first
Content Format	Text	Text, pictures, and short videos (including live broadcasting); text has no more 140-words limitations
Majority of contents	Text	Short videos
Age of Users	Older, majority are adults	Younger, majority are of age between 16–25 (students of middle school, high school and colleges as well as young people graduated from colleges for a few years)
Users geographical distribution	Mostly in tier 1 and 2 cities	More in tier 3, 4, and 5 cities
Drivers for user traffic (the "influencers")	Mostly the "traditional celebrities" (big V), forming an "elite"-centered community	Both "traditional celebrities" (big V) and "grassroots celebrities" (middle V), more decentralized community

By going multimedia, Weibo has improved the quantity and quality of content on its platform—now a typical Weibo sharing is no longer a boring 140-word post; instead, it is a mixture of text, picture, videos, live broadcasting, and interactive comments. With a "mobile first" philosophy, Weibo designed its platform around the capabilities of mobile devices to display content in a simple information feed format. Especially with videos and live broadcasting, Weibo acquired more activities and "stickiness" from its users than in a traditional "social opinions" context. While WeChat has primarily been the place for interactions in smaller circles, Weibo has been the preferred platform for public discussions and sharing.

As a result, the Weibo ecosystem changed fundamentally in three key aspects:

First, it expanded user base by its social media positioning. The main theme of China's mobile internet is about fun and enjoyment. For many people in China, Weibo allows them to be heard publicly and exposed to the rich ideas, cultures, and experiences of the broader world, instead of purely reading about the views from opinion leaders when Weibo was the Chinese version of Twitter. Interestingly, just like in the cases of movies, group purchases, and online literature businesses in this book, Weibo's user growth also came more from the small-town youths than from those in big cities.

Second, it added more active users on the platform (see Figure 4.7). The Weibo platform provides users the means of public self-expression in real time, which is no longer limited to the celebrities. Any user—including the majority who used to be "passive fans"—can create and post a feed and attach multimedia and long-form content. They can also interact among themselves on the platform by following, reposting, adding comments, and sending private messages. During the boom years of 2016–2017, by providing an easy-to-use infrastructure for users to create and share video as well as consume video, especially in a cost-effective manner, Weibo significantly

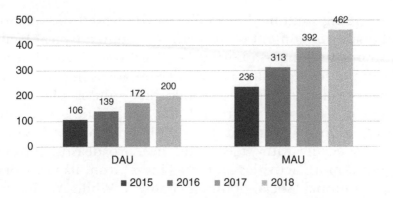

Figure 4.7   Rapid growth of DAUs and MAUs of Weibo (millions, 2015–2018)
Data Source: Weibo SEC filings.

enhanced its user experience: everyone could create UGC (user generated content) videos.

Third, more "internet celebrities" and "e-commerce influencers (KOL)" emerged. Until today, the biggest Weibo stars are still the "traditional celebrities"—the movie stars, sports champions, and business leaders who made their big breakthrough before through TV, cinema, or other traditional channels. Many have tens of millions of followers due to the fact that they are among the first group of celebrities to join the platform since its beginning in 2009 (so-called "Big Vs", who have a "V" behind their name as their accounts have been verified by Weibo).

However, recent years saw the emergence of "internet celebrities" (*Wanghong* in Chinese), who have become self-made online influencers through the internet. These internet celebrities vary from comedians (like the famous Papi Jiang, who became famous by posting funny videos of herself) to fashion bloggers to make-up stylists and even bookworm-like figures who are knowledgeable about esoteric topics. As self-made celebrities, they are also referred to as "grassroots celebrities". The most famous one probably is the fashion influencer Zhang Dayi, whose alleged affair with Alibaba's (once) youngest partner may have changed the CEO succession plan of the e-commerce giant. **(See the "Big V—More than Carnal Interest" box).**

Big V—More than Carnal Interest

Many Chinese online consumers are familiar with Zhang Dayi, a young cyber celebrity who often show up in live-streamed videos wearing a sweet smile and a rimless round-framed pair of glasses while recommending fashion brands. Born in 1988, Zhang began her climb to fame as a fashion magazine model in 2009. In 2014, she set up her own women's clothes shop on Alibaba's Taobao marketplace. She diligently managed her Weibo accounts, whose followers quickly boomed into millions (a "Big V").

In 2015, Alibaba rolled out a program to encourage bloggers, writers, and online experts to post content on various channels on its Mobile Taobao app. Under the program, online experts earn commissions by making product recommendations. Zhang managed the marketing power of her cyber celebrityhood well: on the 2016 Singles' Day, her Taobao shop reached the top 10 of Taobao's Women's Wear category. Furthermore, in 2016, she became a co-founder of the KOL (Key Opinion Leader) agency Ruhnn, which counted Alibaba as an investor and went to the NASDAQ exchange for its IPO in April 2019.

In 2020, more fame came to the fashion influencer. On April 17, a Weibo user believed to be the wife of Jiang Fan, the head Alibaba's Taobao and Tmall online marketplaces, wrote to Zhang Dayi, telling her not to "mess around" with her man. The message quickly caught fire, forcing Jiang to apologize for the outburst from his "family member" and asked the company to launch an investigation into his own behavior. (The Ruhnn shares took a bigger hit, which dropped more than 5% on the day after the incident.)

After looking into the alleged affair, Alibaba said its internal investigation showed that Jiang was not involved in the company's decision to invest in Ruhnn in 2016. (According to public filings, Alibaba's Taobao had an 8.56% stake in Ruhnn.) Neither preferential policies toward Ruhnn nor Zhang Dayi's Taobao and Tmall stores were discovered during the investigation, Alibaba said. However, Alibaba announced that while there had been no related transactions, Jiang had hurt the company's reputation with his "improper handling of family matters".

Only 35 years old, Jiang was the (former) youngest member of Alibaba's partners' committee. Leading the largest e-commerce platform of China, he was widely seen as the next generation CEO of the Alibaba Group. After the investigation, Alibaba removed Jiang Fan from its partners committee, demoted him from senior vice president to vice president, and clawed back his 2019 annual bonus for the misconduct. For Alibaba, this could mean the time for a new CEO succession plan.

The grassroots celebrities may have millions of followers or hundreds of thousands of them, and they can be categorized accordingly as "Middle Vs" and "Small Vs" (vis-à-vis the "Big Vs"). What's significant about the Middle V and Small V players is that, just like the "Middle Gods" and "Small Gods" in the online literature businesses, they form the core and the most active part of the ecosystem. With sizable (but not of dominant number) fans following them, they aggressively monetize their social assets on Weibo through advertising, e-commerce, subscription, tipping, and other means, leading to ever-innovative synergies between entertainment content and e-commerce.

As a groundbreaking example of the marketing power of grassroots celebrities, in April 2016, an online-video star named Papi Jiang, known for her sarcastic social commentary on her Weibo page, auctioned her services as a spokes model to the highest corporate bidder. The winning bid—at US$3.4 million—came from Chinese cosmetics retailer Lily & Beauty, which received a one-time promotion during one of Papi Jiang's videos and a mention on her social media accounts. In another example, fitness expert Chen Nuanyang aired a live video broadcast each week on her Weibo page, and fans logged on to learn as much about abdominal crunches and yoga poses as they did her latest line of workout togs. She leveraged an online base of 750,000 Weibo followers into a Taobao store that produced more than US$2 million monthly workout gear sales in 2016.

In summary, Weibo's successful move was to shift from text micro-blogging to multimedia and truing its traditionally passive fans into active content creators. However, users' appetites and demands for entertainment are growing and insatiable. Since 2018, the market sees a craze in China for the new genre of short video apps. As a result (among other market factors), Weibo's user growth relatively slowed and stock price experienced pressure of late (see Figure 4.6). As of early 2020, Webio had just over 500 million monthly active users (compared to Twitter's 300 million), representing slowing user increase since

2018, when new players like TikTok (Douyin) started to take away user traffic from Weibo by providing fans with trendy video tools and "addictive" recommendations.

TikTok: Goofy Videos Dominating Social Media

Short-form videos, a genre of quick-hit entertainment, have seen rapid growth in China in recent years. First, quirky and fun mini-clips provide a different viewing experience from full-length videos and movies. On subways or at shopping malls, there are always people viewing short videos on their smartphone screen when they have a few minutes (or seconds) of time interval. Second, the younger generation are increasingly turning to short videos as the mean for social interaction. By providing an easy-to-use infrastructure for users to create, share, and consume short videos, Douyin (TikTok is its international name) and Kuaishou have become the two leading short video platforms in China.

Literally translated as "shaking voice" or "vibrant voice", Douyin was launched in 2016 by Beijing-based software company ByteDance, which also controls Toutiao (or "Today's headlines"), China's most popular news aggregator. This group (see Figure 4.8) is usually referred to as *Toutiao* in China, which is the *T* in TMD, the acronym for the three new tech leaders after BAT (see the related discussion in Chapter 2). ByteDance

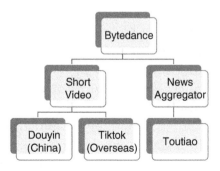

Figure 4.8   The TikTok family

is valued at more than US$100 billion since 2020, making it the world's biggest unlisted media startup. As such, this section focuses on Douyin/TikTok.

In 2017, TikTok acquired the teen lip-sync karaoke app Musical.ly in the US market for US$1 billion. In 2018, ByteDance folded Musical.ly into TikTok and rebranded it to a single application under the TikTok name for its overseas markets. It took off in the United States, as well as large emerging markets like India and Southeast Asia. According to the Sensor Tower report that ranks the top 20 most downloaded apps of 2019, Facebook owns four out of top five apps: WhatsApp, Messenger, Facebook, and Instagram, but TikTok is the second most popular right after Whatsapp (and ahead of YouTube, Snapchat, and Netflix) after TikTok/Douyin amassed a combined 740 million downloads in 2019.

In the United States, TikTok has attracted large number of users from diverse communities. One of the active groups are young military service members. They upload videos of themselves, often executing fitness exercises in uniform inside military facilities, often looking like a war theater. As the COVID-19 pandemic has forced the world's teenagers out of school and into their bedrooms, they have turned to TikTok for silly videos. (According to news media, the app was downloaded 115 million times in March 2020 alone.) Such popularity, plus the enormous data it accumulates from the vast user base, has alerted the US government regarding national security and data privacy issues, which will be discussed in Chapter 9.

What's the magic in TikTok? Basically, TikTok is a short-form video app with powerful editing capabilities such that all the ordinary users can participate with their talents and creativity. It enables users to create 15-second videos with additional special effects, such as shaking and shimmering colors and popular hip-hop dance music, before sharing it around. The special attraction is that TikTok has more polished, better quality, and more smoothly crafted content than other short video platforms in China. Most of the videos are produced by adolescents doing nothing fancy—acting, dancing, doing gymnastics, lip-syncing, and performing silly stunts. Yet, thanks to digital technologies, they form a distinctive meme: a dare, a prank, a teenager looking pretty.

For example, during its early growth phase in 2017, the "Karma's a bitch" makeover videos, which featured users transforming their appearance in a flash while lip-syncing to "Karma's a bitch," went especially viral. The "Karma's a bitch" challenge plays into TikTok's special effects capabilities perfectly. The user looks into the camera, having cool music playing in the background, lip-syncing to Veronica Lodge uttering the "karma is a bitch" line in the US teen soap *Riverdale*, throwing a scarf or sheet up to cover his or her face, then letting it drop to reveal the better (or best) self in a great outfit or with perfect makeup. The users must feel empowered by TikTok to look glorious and incredible before his or her enemies. *Don't be bitter, get better*, right?

The *get better* aspect of TikTok is the core strength of this digital tech-centered platform. To start with, creating interesting video content is really challenging. Adding the fact that the Chinese are often known for being shy and reserved, many users would hesitate to perform and create their own videos without professional support. For that, TikTok provides enough video templates, guidance, and examples to ensure users' interest in creating their own videos.

On top of that, TikTok makes you look even *better*. TikTok offers a wide range of (up-to-date and always changing, as required by the teens) background music and special effects to make the videos richer and funnier, including stickers, change of appearance, slow motion, frames, and more. All the ordinary users can try out the seemingly endless possibilities to create professional-looking and highly polished videos (PUG Videos). With the digital tools of the app, everybody—even the most talentless dabbler—becomes a talented performer with a cool sense of humor.

In another example, in the 2018 FIFA World Cup, China's national football team (again) failed in the qualifying round, missing the final tournaments in Russia. Yet the mobile internet nevertheless helped to create a festival in China for hundreds of millions of fans as well as first-time watchers. During the World Cup, they created their own videos recording their lives during the tournament. Lots of girls—who were not necessarily football fans—also feverishly joined this sports festival with creative videos of their own. Thanks to TikTok, average-looking girls turned themselves into sexy "football idols" like professional models.

What's the most powerful about TikTok is the rapid emergence of key opinion leaders (KOL, akin to the "influencers" who are popular on Instagram and YouTube) on its platform. The platform uses sophisticated machine-learning algorithms to facilitate exponential audience growth and create KOLs out of users. In fact, it may surprise many about how easily new users can reach a vast audience in TikTok. Here is why:

On the one side, TikTok applies a powerful recommendation AI algorithm to learn users' interests from their viewing habits. The algorithm is designed to track content browsing and interaction histories to analyze the correlation among video content, user characteristics, and user behaviors. When a new user first opens the app, the initial hundred videos shown to him or her are more or less random and standardized. All videos replay by default, which often leads to multiple re-watches. This short video format trains the algorithm very fast: in the time it takes to watch a 15-minute long-format video, TikTok can capture user behavior data from 100 videos of 15-seconds (including the replays, which has heavy weighting in the algorithm). Once the algorithm has gauged a user's preferences, it feeds them more targeted material from its content pool.

What's next is extremely simple and effective: TikTok bombards the "known" user with self-repeating clips. It continuously serves similar lighthearted videos people didn't even know they wanted until the machine-learning algorithms pushed them, in an increasingly personalized stream, until users are hooked. After a few rounds of feeding and feedback, the AI algorithm can be so effective that TikTok is called as the most "addictive" app in China because it's become one of people's favorite ways of killing free time. In the Western markets, TikTok is referred to as "YouTube on steroids".

On the other side, TikTok consciously channels the best videos—determined by their popularity—to the most viewers, thus making the winners even bigger winners. When a first-time video is greeted by a small number of selected audiences positively (they praise, like, or simply wait out the full clip), the algorithm recognizes their approval and distributes the video to a larger group (also based on the estimated interest by the system). This process is repeated if more viewers continue to like the content. With the two sides of algorithm power multiplying, for users with popular content, the system allows them to quickly grow their fan base and become KOLs.

On top of the algorithm, TikTok also puts in human capital to cultivate rising KOLs. According to a news story of the *Wall Street Journal* in April 2020, Holly Grace, a 23-year-old part-time singer and full-time nurse in Nashville, Tennessee, started making videos for TikTok by lip-syncing to whatever 15-second snippet of pop music is trending on the app every week, and she would also put her own twist on other viral TikTok videos, usually referencing topics relatable to medical professionals. When she had about 400 000 followers, completely unexpected by the young nurse, TikTok reached out to her and gave her a "manager" within the company to help make her videos even more viral by flagging trending songs and discussing what had and hadn't worked in her past videos.

Going forward, TikTok (the ByteDance group) is looking to ride the TikTok momentum and expand into e-commerce, games, and other areas as the company tries to compete against tech giants globally. For example, Jiaqi Li was a Maybelline counter beauty adviser in Jiangxi before becoming one of Douyin's most popular male celebrities and the best lipstick salesman. Li gives females recommendations about how to choose lipsticks by personally wearing the different shades and has become the current keeper of the Guinness World Record for applying lipstick on the most models in 30 seconds. Once he sold 15 000 lipsticks within 15 minutes of livestreaming (a striking number compared with a few hundreds the makeup artist could do per day at outlets), beating Alibaba founder Jack Ma, who wanted to compete with Li in a two-hour livestreaming session selling lipsticks.

Further, the platform also expands into more media format. Few fans of cheeky videos have noticed that ByteDance is also investing heavily in online reading. ByteDance announced to ramp up its online literature business by integrating all the novels published on Toutiao app into Tomato Novel, its own e-reading app. On the 2020 World Book Day, ByteDance had its contracted authors doing live streams on Douyin and TikTok to promote their works.

Meanwhile, *globalization* has been a key word of Douyin/Tiktok's strategy after its huge success in China. Unlike its peers, ByteDance does not merge Chinese and international coverages; instead, it creates a separate app for TikTok specifically for going abroad, which proves to be an extremely savvy (yet still insufficient) move amid the US–China tech war.

For one thing, Chinese internet culture does not always translate in a global context. Even for the super-app WeChat messaging that claims more than one billion users worldwide, it still has little impact in the English language-based social landscape dominated by Facebook or in local-language-centered markets such as Japan. By contrast, Douyin and TikTok are branded as the same product, but with distinct characteristics for different marketing targets. TikTok worked well in the overseas markets, and as such, in 2020, ByteDance is the only technology firm bar Apple with more than 100 million users both in the Chinese and Western markets.

For another, the separate business setup allows TikTok to be managed "locally" in the foreign markets. In May 2020,

Disney's Head of Streaming, Kevin Mayer, was recruited to become TikTok's new CEO. This move added to TikTok's continuous efforts to distance itself from China, but was still short of appeasing the US government, which is wary of China's rise in the cybersphere.

In August 2020, US President Trump issued an executive order whereby ByteDance must divest its US operations of TikTok within 90 days, and Meyer soon resigned from his new job. "In recent weeks, as the political environment has sharply changed, I have done significant reflection on what the corporate structural changes will require, and what it means for the global role I signed up for." Meyer wrote in a letter to employees. Many Chinese tech companies have tried and failed to work out their strategy beyond the home market, and it seems that TikTok/ByteDance is unlikely to be the first global digital media company that can succeed both inside and outside of China. (See more discussions relating to China–US tech war in Chapter 9.)

Kuaishou, iQIYI, and More: Into the Multiple Screen Competition

In concluding this chapter, the case studies have all illustrated the key components of successful fans economy ecosystems: the bigger number of content creators, the more diverse and interesting contents are supplied, the more active and sustainable the platforms become (see Figure 4.9). While the platform should focus on the celebrities at the top of the pyramid, a successful content ecosystem can only be as strong as its middle core, which includes both avid fans and active content generators themselves—whether the Middle Gods of online novels, Middle V grassroots celebrities of Weibo, popular KOLs of Douyin, or avid uploaders of Bilibili.

In order to cultivate, engage, and motivate the "core", the platforms need to not only amass the largest possible user base for media contents consumption and sharing, but also provide video production tools and monetization means to incentivize

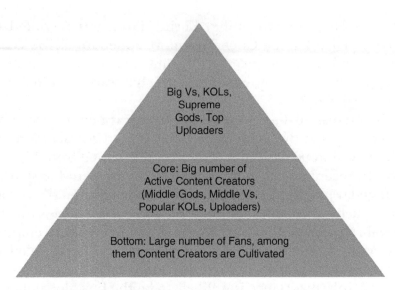

Figure 4.9   The ecosystem of the fan economy

ordinary users to become the most creative PUGC content generators themselves. That's no easy task. The current leading player, TikTok of ByteDance, is the perfect example of the sustainability issue. With few voiceovers, the 15-second Douyin/TikTok content easily travels globally. For the same reason, however, there is no real network among the users themselves. Plus, how much advertisement could be added to a 15-second clip? And money matters: ByteDance still generates most of its revenue from China, despite the TikTok app's rapid international expansion.

In fact, competition is heating up because the online entertainment competition is far from over. It is facing challenges from both startup rivals in the same field and established tech giants that are starting to embrace short-form content. Short video has been popularized by TikTok for sure, but the barrier to entry is relatively low. Attention is a precious commodity for the Generation Z users, who could easily switch to a new, "cooler" short video platform as they jumped on TikTok in the first place. Furthermore, TikTok's "addictive" recommendation

AI seems to work effectively with short-form contents, but can it be the same effective with longer-format contents?

In China's online entertainment market, the competition to become the ultimate one-stop entertainment platform is fierce. TikTok faces significant competition from many companies that offer various content to users and captures their time spent on mobile devices.

First is **Kuaishou,** Douyin (Tik-Tok)'s direct competitor in short-form videos. According to an early 2020 media story, Kuaishou reached 300 million DAU (daily active users), showing a strong year-on-year growth as well. It is the preferred video platform in less developed areas across China; therefore, it may enjoy a big user base from rural, less affluent people as the mobile internet continues connecting more first-time internet users. One big difference from TikTok is that Kuaishou's platform is organized by both personal relationship and machine recommendation, whereas TikTok is mostly about machine recommmendation. (If YouTube is the closest to TikTok, Kuaishou shares some similarity with Snapchat Stories.)

The Kuaishou's platform values PUG videos that attract viewers' attention, "likes", and interactive comments. Therefore, for the most popular videos, the viewers not only have interests in the fun content but also the video creators themselves. That difference is evident in e-commerce video streaming, where Kuaishou celebrities' income is generally higher than TikTok KOLs. The personal trust and emotional attachment in the Kuaishou relationship could be valuable in the long term. Furthermore, in 2020, Kuaishou acquired the bullet screen site Acfun, the earlier competitor to Bilibili. (For people who cannot understand their trendy names, the two used to be called "A site" and "B site".) It could be really powerful if Kuaishou could integrate short video, bullet chat, and more Generation Z features into one platform.

Second, the group of long-form video platforms, led by **iQIYI** (controlled by Baidu), **Youku** (acquired by Alibaba), and **Tencent Video** had their glorious years before (see the author's 2016 book *China's Mobile Economy*), but they have consistently suffered operation loss and still have no clear path to profitability. The problem is they have the same business model, which is to provide free content to attract online viewer's eyeballs and rely on advertising income as the main revenue source. Just like YouTube in the United States, the Chinese sites have provided viewers the access—largely free—to nearly all the latest Chinese TV shows, movies, documentaries, professional chess games, and so on.

In the end, all the sites are converging into similar content offerings, and in the case of hit variety shows, TV series, or movies, the content from all video providers are the same. So if one streaming site updates the content too slowly or stops updating, the viewers can simply move to another site (and move among different sites back and forth) since the cost for switching between the sites (essentially zero) is much lower than switching cable TV services. Due to the similar content available on multiple platforms, it's challenging for the video sites to develop a stable user group, which is critical for both subscriber fees and advertising income, which in turn, puts further pressure on the video streaming sites to spend more on acquiring "unique" quality content, driving up operation costs.

But that does not mean long-form videos are out the window. People still want good stories, and it is hard to tell a (good) story in *15 seconds*. And what's the appropriate advertisement time for such a short video? (In fact, TikTok recently extended its video time limit to 15 *minutes*). Advertisers still favor long-form videos like drama series or game shows, where the IP library of established PGC video platforms is a meaningful resource.

Now the long-form video platforms are also betting on more user engagements and PUG videos. In 2019, iQIYI, Youku, and Tencent Video all launched interactive video products, so that audiences are no longer passive viewers and can make choices for the characters and changing the story as it goes along. iQIYI launched the world's first professional Interactive Video

Guideline to standardize interactive content creation in May 2019, and other platforms quickly followed. Using the platform tools, creators can produce videos that let viewers decide what a character does to experience different endings.

In the first interactive film and television work in China, *His Smile*, iQIYI gives audiences more choices for a romantic comedy. The video is about a young talent manager at her new job overseeing members of a rising boy band looking to make its big break. The story is told from her perspective as romance inevitably ensues, and the audience—especially the young girls—surely have different views about "picking the right one". The "branching plot" function enables the viewers to choose their own adventure through 21 different branching story options and 17 possible endings. When the technology matures, the convergence of long- and short-form videos can be an interesting new direction.

Last, and certainly not the least, are the BAT giants, especially Tencent of the entertainment DNA. Alibaba, Baidu, and Tencent are all tapping short-form video to pull young users to their platforms, but Tencent's DNA is in online games and social media; hence, it takes on ByteDance head-on. Tencent has blocked most Douyin videos from its ubiquitous app WeChat (apparently out of concerns for content compliance in China), and the two rivals (the two founders, not their staff) have swapped online accusations of anti-competitive behavior. (In the earlier online literature section, Tencent made key changes to its senior management team at China Literature in 2020 and shifted its business model from subscription-based to a free content and ad-based model. Apparently, Tencent was following what ByteDance is doing with online literature.)

All in all, the ultimate competition is for the time and attention of Chinese internet users. Everyone aims to create a one-stop entertainment platform that match users with all possible entertainment options, so they do not have to leave the platform when they want to read books or play games, create videos, or share anime. The rising challenge is that the industry

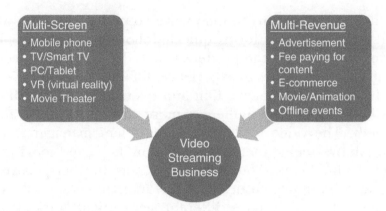

Figure 4.10   The coming age of multi-screen, multi-revenue model

may have reached the "peak screen time"—in fact, the rise of short videos probably reflects the decreasing attention span of ordinary users when they are online. As a result, users are expensive to recruit and difficult to keep.

Therefore, the internet firms must capture and keep users' continuous attention no matter what screen they are watching, online or offline (the movie cinemas)—and convince them to pay. As they compete to offer the most compelling content package, terminal devices, and integrated platform, the "multi-screen, multi-revenue" model is the new game in town (see Figure 4.10). The future entertainment content, based on the upcoming 5G network, would be produced into various form of content, which then could be seamlessly watched across multiple screens of a smartphone, PC computer, smart TV, movie theater, and more—and the competition among platforms will be more complex and entertaining.

5

The Heartland of Blockchain and Fintech

- The Heart of the Bitcoin and Blockchain Network
- Red Envelopes and Mobile Payments
- Crowd-funding and P2P Lending
- Micro Loans, Big Data Solutions
- AI and Alt-data for Wealth Management
- Blockchain and SCF: The alliance of Alibaba and ICBC

The Heart of Bitcoin and Blockchain Network

Chinese entrepreneurs have been on the crest of the blockchain wave since Bitcoin first gained traction. China is home to some of the world's largest farms"—data centers hosting the highly powered computers, where the so-called Bitcoin miners (companies or passionate individuals) compete against others in the blockchain network to solve complex math puzzles and earn new coins.

Bitcoin miners draw on huge amounts of computing power as they battle against others to produce hashes—alphanumeric strings of a fixed length that are calculated from data of an arbitrary length. (That power is known as the "hashrate", and it dictates a computer's ability to produce new coins). To lower the related costs, the computers are often hooked up to cheap, plentiful electricity by the mining companies and individual miners. In short, electricity usage and semiconductor engineering are two important factors for mining costs, which have a significant impact on the investment return of Bitcoin mining.

First, China is a large country, and there are still many underdeveloped areas with low-price, abundant electricity, such as Sichuan and Yunnan in Southwestern China that have an ample supply of cheap hydroelectricity as well as coal-rich Xinjiang and Inner Mongolia in Northwestern China. The Bitcoin boom used to be driven mainly by dirty coal, but more recently, Chinese miners have deployed more machines in Sichuan and Yunnan provinces to take advantage of the even cheaper hydropower—especially during the rainy season. According to CoinShare's estimate in December 2019, approximately two-thirds of global Bitcoin mining (65%) happens in China, and that Sichuan province alone produces more than half (54%) of global hashrate.

Second, China manufactures most of the world's mining equipment. Chinese companies such as Bitmain, Canaan, and MicroBT are among the world's biggest manufacturers of bitcoin mining gear. They sell Bitcoin mining rigs—the specialized computers that solve complex mathematical equations to earn or "mine" new Bitcoins for their owners. (CEO of Bitmain once claimed that globally 70% of the Bitcoin mining rigs in operation were made by his company). CoinShare's December 2019 estimate showed China's bitcoin miners controlling 65% of the crypto network's processing power, up from 60% in June 2019, was the highest recorded by CoinShares since it began tracking hashrate. The gains may be due to China's greater deployment of more advanced mining hardware, according to industry analyses.

Another Bitcoin information website, buybitcoinworld-wide.com, in 2019 estimated that Chinese mining pools even controlled more than 70% of the Bitcoin network's collective

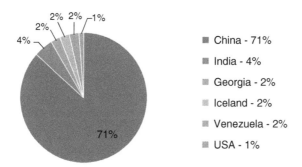

China - 71%

India - 4%

Georgia - 2%

Iceland - 2%

Venezuela - 2%

USA - 1%

Figure 5.1   China dominates Bitcoin mining's hash power

hashrate. In its hash power breakdown by country, the web-site assigned 71% to China (see Figure 5.1). 15% of the hash rate is missing from the chart, but it's likely, suggested by the website, that China controls an even greater amount. By all means, China is the undisputed world leader in Bitcoin mining. If a couple of big Chinese miners were to switch off, the hashrate will likely drop sharply, making the network less robust and secure.

Third, China also accounts for hefty Bitcoin trading volumes. The Chinese exchanges for cryptocurrencies (Bitcoins and other tokens) used to lead the world in terms of volume. The "coin" rush was interrupted in 2017 by Chinese authorities, which effectively banned all Initial Coin Offerings (ICO) activity within China as "unauthorized and illegal public fundraising" and "unauthorized public sales of securities". The Chinese government also made illegal all cryptocurrency exchanges within the country. As a result, the market saw cryptocurrency trading and other related activities in China moving abroad due to tightened regulation. (E.g. in early 2020, Bitmain was reportedly building the world's largest bitcoin mining farm in Texas, which also has abundant power resources.)

In contrast to cryptocurrency (which touches on the country's "digital currency", see the related discussions in Chapter 1), the Chinese government endorses blockchain as "a core technology". After the initial Bitcoin and then Ethereum (smart contract)-based tokens, since 2018, the blockchain technology has reached its third phase and been integrated with the real economy. For example, in the fintech fields, blockchain is used by banks to revolutionize the supply chain financing model, which are discussed in this chapter. This trend is powering forward after President Xi Jinping's public remarks in late 2019, urging more efforts to accelerate the development of blockchain technology and industrial innovation.

As mentioned at the beginning of this book (see Figure 1.1), Chinese firms also dominate blockchain patent filings globally, and they are developing practical applications for various industries. At the top of the patent ranking is Alibaba and Ant Group, the financial arm of Alibaba. The blockchain prowess of Ant Group, however, it's just a small part of its empire (see Figure 5.2). Its e-payment division, Alipay, for example, processes more than half of the volume in the largest mobile payment market of the world (Apple Pay, by comparison, has less than 1% market share). According to research firm Analysys, Alipay in 2019 handled more than US$1 trillion worth of payments.

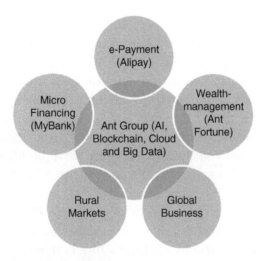

Figure 5.2   The fintech empire of Ant

Ant Group also compiles personal credit scores based on Big Data, and consequently, by offering as little as a few bucks at a time to new borrowers, its microlending business has quietly swelled into one of China's largest providers of personal credit lines. Furthermore, it becomes a powerhouse for wealth management, not only providing digital distribution of financial products, but also developing new AI-based trading tools for asset managers. Its broad array of services is indispensable for China's digital middle class, and its reach is expanding into rural China and global markets.

That's why Ant Group is the most valuable private fintech firm in the world (actually, the biggest among all tech unicorns). In 2018, Ant Group raised around US$14 billion in what market watchers called the biggest-ever single fundraising globally by a private company (according to market data firm Crunchbase, US$14 billion amounted to the largest confirmed single fundraising round in history). The imputed valuation US$150 billion (in Chinese RMB/yuan currency, over 1 trillion) was greater than the market capitalization of Goldman Sachs, the famous US investment bank. In August 2020, it filed for a joint listing in Hong Kong and Shanghai, targeting a valuation of more than $200bn.

In short, there is no equivalent to China in the fintech world today. As internet companies transform the smartphones into platforms for financial transactions, China is far ahead of the rest of the world in terms of how widely internet finance is used. For a large part of the Chinese population that has never even used credit cards, they are now using the internet to manage their payments, savings, and investments, just as in many parts of China, people are skipping landline phones in favor of a smartphone to access the internet for the first time in their lives.

In addition to Ant Group, its internet firm peers like Tencent and traditional banks like ICBC (the Industrial and Commercial Bank of China) all have become fintech giants. Their AI research and inclusive blockchain infrastructures are transforming the banking, insurance, and wealth management practices. Before going into the latest developments, this chapter starts with mobile payments, the origin of all fintech revolution as well as the internet-based financing models (equity crowd-funding and P2P lending), which had a dramatic rise (and fall) last few years, as a background.

Red Envelopes and Mobile Payments

During the Spring Festival (the lunar New Year), it is a centuries-old custom for Chinese families to hand out red envelopes (called *hongbao*) stuffed with crisp money bills to relatives and friends to bid good fortune. The traditional *hongbao* scene used to be children in red holiday dresses kneeling down and kowtowing to their elders to receive the gift, but now in the digital era, it has become a fun game on mobile devices.

The red envelop was first introduced by Tencent at the eve of the year of the Horse (early 2014) on its popular messaging and social network platform WeChat. Individuals sending greetings to families and friends, and corporations showing appreciation to its employees, could conveniently do so by sending digital red envelopes from WeChat. The *hongbao* giver only needed to link the digital *hongbao* function with his or her payment channel (a bank account, e.g.). The giver could then directly transfer *hongbao* money from bank accounts to the receiver through Tencent's payment system on WeChat ("WeChat Pay"). On the receiving end, the receivers also had to link their bank accounts to get their *hongbao* in cash or use the digital cash to pay for e-commerce services and products, such as mobile taxi-hailing (see Figure 5.3).

In addition to the digital red envelopes sent by the WeChat users, Tencent itself also sent out a large number of *hongbaos* freely to the public. Of course, Tencent was not just playing the "Santa" role for Chinese netizens' festival celebration. The animated digital envelopes were incentives for users to experience its mobile payment system. Because both the givers and

Figure 5.3   The parties in a red envelope—fun and value

receivers had to install the app linked to the WeChat Pay system, as well as to register their banking accounts or debit cards, Tencent was hoping that the netizens would happily turn themselves into WeChat Pay users while playing a fun game.

To the surprise of many internet firms (including Tencent itself), the concept of digital red envelopes falling from the internet received groundbreaking acceptance from Chinese customers overnight. Many found gifting through electronic services more convenient and safer (versus traveling with cash). Young people, in particular, liked the "fun" aspect. Because the number of free red envelopes given out was limited, people had to rush to the links to secure a *hongbao*, and this sensational process was referred to as "grab (or fight for) red envelopes"! in Chinese. The "grab" feature turned a passive custom into a participative activity and added an element of suspense, making the process more fun for the users to play.

The fun features of WeChat red envelopes created an extremely successful marking campaign for WeChat Pay (also known as "Tenpay"). With limited cash input, Tencent had multiplied active users, acquired important data (banking information), and familiarized users with its WeChat payment service. During the 2014 Lunar New Year season, hundreds of millions of users were glued to their smartphones instead of holiday activities, trying to "grab" as much as possible the virtual red envelope money being doled out by their relatives, friends, and Tencent itself. As a result, people opened the WeChat Pay app more often than they opened the Alipay app, even though WeChat only launched the payment service about 10 years after Alibaba started Alipay.

Launched by the Alibaba Group in 2004, Alipay has the longest history in China's third-party payment market. After the restructuring around its 2014 IPO, Alipay is not owned by Alibaba, nor is part of the Alibaba IPO, but it is controlled by Alibaba's top executives. Alipay runs a PayPal-like internet-payment service and it is already the world's No. 1 processor of mobile payments. Alipay not only processes payments on Alibaba's Taobao and Tmall online marketplaces, but also handles many other types of online and offline payments, including utility bills, ticket booking, restaurant meals, and taxi rides.

However, built on top of the explosive penetration of the Wechat app, WeChat Pay showed strong momentum of growth in recent years. After the innovative digital red envelop campaign in the 2014 Chinese New Year, WeChat Pay successfully made its major push into the online payment business territory long held by Alipay. The digital red envelope instantly attracted millions of WeChat users to sign up for its payment service for the first time, based on which Tencent doubled its market share from 5% to 10% within just a year. By the end of 2018, Tencent's mobile payment system already reached 38.87%, rapidly catching up Alipay's 53.78% (collectively, they control more than 90% of the market). Apple pay, by contrast, did not even make the top 8 list according to Analysys research report (see Figure 5.4).

Looking beyond the New Year's fun and drama, the war for red envelopes was a serious fight between the two internet giants for the future of the mobile internet. To start, mobile payment service is a critical component of mobile e-retailing, which aligns with the Chinese consumer's desire for speed and convenience of "any time" shopping. Many of these purchases are ad hoc decisions arising from offline world advertisements, social conversations, or simply random thoughts. As a result, a large percentage of purchase decisions may not eventually turn into actual purchase transactions if the e-retailing platform

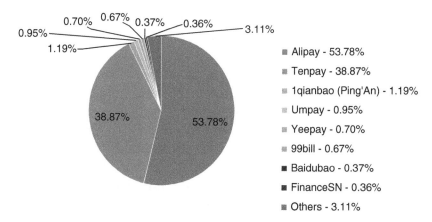

0.70% 0.67% 0.37% 0.36%
0.95%
1.19%
3.11%

- Alipay - 53.78%
- Tenpay - 38.87%
- 1qianbao (Ping'An) - 1.19%
- Umpay - 0.95%
- Yeepay - 0.70%
- 99bill - 0.67%
- Baidubao - 0.37%
- FinanceSN - 0.36%
- Others - 3.11%

38.87%   53.78%

Figure 5.4   Mobile payment market share breakdown in China
Data Source: Analysys Report, December 2018.

does not have a ready and convenient mobile payment function attached to it.

For Alibaba, that was a stronghold that it could not afford to lose. Comparing to Tencent's WeChat Pay, Alibaba's Alipay had a longer history and a head start with hundreds of millions of active users from its affiliation with Taobao, TMall, and other Alibaba's e-commerce sites. For Tencent, Wechat, the most popular social forum in China, is an important platform for brands (especially for foreign) to develop multimedia marketing campaign (the corporate accounts); meanwhile, individual WeChat users could establish in-app stores (so-called "little WeChat stores", *Weixin Xiaodian* in Chinese) to sell goods to their friends and followers. By integrating mobile payment function, Tencent, having major brands to open stores inside WeChat, is a threat to Alibaba's Tmall, and its inclusion of smaller merchants is a challenge to Alibaba's Taobao.

Overall, mobile payment is much more widely used in China then in developed markets like the United States or Europe. The prominence of digital payment systems isn't just limited to upscale, exclusive services—even food stalls and fruit vendors (and street beggars)—have been heavily embracing digital payment systems. In addition to digital retailing, the mobile payment system has become an important link between mobile users and their offline spending activities, such as restaurant booking, grocery deliveries, in-home manicures, and movie ticket ordering, just to name a few. Many of those activities happen during the "fragmented time" of users when they are "in between" the online and offline worlds. (See Figure 5.5 and more O2O business discussion in relating to the "shared economy" in Chapter 6.)

Since 2019–2020, the state-owned China UnionPay Co. Ltd. (**UnionPay**), which runs the country's banking card monopoly (the Chinese counterpart of Visa or Mastercard), has achieved rapid growth in its mobile payment network. According to media reports, it reached 240 million users by March 2020, becoming the one most likely to disrupt the mobile payment duopoly. (According to the data from Analysys, Alipay and WeChat Pay still control more than 90% of China's mobile payment market.) In early 2020, *Caixin* magazine reported that Tencent has agreed to work with UnionPay to integrate the QR code systems in their mobile payment platforms. This collaboration will expand UnionPay's market penetration and further China's central bank's plan on an integrated payment system across the digital space. (Since 2017, Alipay and WeChat pay have already been required to route payment transfers to and from banks through a government-backed clearing system.)

Figure 5.5   "Fragmented time" and O2O activities

Going forward, all the major payment players are focusing on the less developed "to B" payment businesses, such as global trades and logistics. Just like the two tech giants, Union Pay has increases investments into blockchain services, including a digital certificate application and a blockchain-based tracking platform for cross-border capital transfers. Since Unionpay is also leading the Blockchain Standard Network (BSN) platform that links both domestic and international markets (see the detailed discussion in Chapter 10), it will be a strong competitor to Alipay and WeChat Pay in many areas of digital payments.

Crowd-funding and P2P Lending

In a short period of time after the spread of mobile payments, internet finance has fostered a more inclusive financial system in which loans and equity financing are more widely available to consumers, startups, SMEs (small- and medium-sized enterprises), and whoever can put them to productive use. New players, many of which are private, internet-based, and light on assets, have entered China's mostly state-controlled financial industry and offered innovative products and services. However, due to the lack of market entrance threshold or business operation standards, many of these new offerings have evolved from regulatory "gray zones" and have grown rapidly "in the wilderness".

The rising integration of internet and finance—almost to the extent of out-of-control before stringent regulations were put in, as illustrated by the following examples, is closely linked to two imperfections of the financial system—and potential solutions from the internet. **One** is the lack of investment choices for individual savers and investors. China's financial services sector is immature compared with developed markets, and it lacks the variety of products or services found in the US market. The two major investment channels—public stock and real estate markets—require large investments and a high risk tolerance; meanwhile, the deposit interest rates at banks are tightly regulated and currently set at low levels.

The other one is the difficult access to credit by individuals or small and early businesses because the banks focus their businesses on the bigger and established companies for their perceived lower credit risk. In response, new lending and equity financing models are initiated for the under-served individuals and SMEs, such as equity-based crowd-funding and peer-to-peer (P2P) lending.

Crowd-funding is a capital-raising transaction from a large number of people, with each person contributing only a small amount to the whole project. At the beginning, crowd-funding was considered to be an equity financing channel for small- and medium-sized enterprises (SMEs). In China, private companies must seek government regulators' approvals before listing, and the qualification requirements for public issuance of equity are highly rigorous. It is really challenging for SMEs and startups to be eligible for the public issue of securities, and the costs for the listing process and the continuous disclosure filing requirements are prohibitively expensive for SMEs.

However, the important legal issue is whether a crowd-funding transaction constitutes a public issuance of equity, which is regulated by China Securities Regulatory Commission (CSRC), or it can be structured to be a private placement exempted from registration requirements. After a short while of market confusion, the CSRC and other government ministries have formally determined that a crowd-funding of corporate equity characterized by "public distribution, small size, and wide participation" should be treated as public offering of equity securities and subject to regulation (see Figure 5.6 for different standards for "public" and "private" crowd-funding transactions).

As such, crowd-funding has failed to be an important financing channel for corporations. Nevertheless, the concept of crowd-funding has quickly become well-known in public because it plays into the "fun" theme of the mobile internet.

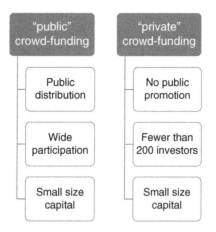

Figure 5.6   Crowd-funding: Public offering versus private placement

Most crowd-funding projects focus on small investments for hit movies, unique experience, and social entrepreneurship. With little investment involved, the downside risk for investors is limited, and in the least, they may have some fun and enjoy some interesting stories in return.

For example, crowd-funding has changed how films are financed in China. Traditionally, Chinese films have been funded by studios, either state-owned or private, including the new ones set up by the internet firms themselves. In the digital age, the crowd-funding campaigns are marketed as an opportunity for ordinary audiences to get involved in the glamour of movie making since anyone who loves films or adores stars could become a minor investor and participate in movie production and promotion. As the "mini-producers", the investors like their chances of visiting the filming locations, getting the exclusively issued electronic magazines, and even meeting the stars.

Internet giants, including Baidu and Alibaba, have all launched movie financing products, where users could put up as little as 100 RMB (US$15) of their own money and be able to claim themselves as an investor of the latest blockbuster movies. They were excited to be "co-producers" of the movie, and appreciated the chance to visit filming locations, receive stars' autographed photos, and even meet the stars. In a way to gather information on audiences' taste preferences, Alibaba's movie investment product covered a range of projects from movies, TV shows, and video games, and each investor was asked to choose a maximum of two investment plans from the list of projects. For the internet firms, in addition to further familiarize users with internet finance, they also build up data on moviegoers' taste preferences.

Online peer-to-peer (**P2P**) lending is the practice of ordinary consumers lend directly to individuals and small businesses over online platforms, without using a traditional financial intermediary such as a bank. While crowd-funding attracts many *diao-si* investors to make a small investment for fun, P2P lending appeals to a smaller group of wealthy investors who are far more willing to take risks in large amounts.

The P2P lending business is part of the shadow banking market in China, as many P2P lending websites gather funds from public and then lend funds to individuals or small companies, instead of simply serving as an information platform to facilitate lending and borrowing between parties. It emerged as a valuable source of credit for consumers and small businesses that cannot access loans through the traditional banking system. As a result, P2P exploded in popularity over the last few years in China (see Figure 5.7), and China quickly surpassed the level in the United States (where the model was first developed) and became the world's largest peer-to-peer lending market since 2015.

The essence of "peer-to-peer" lending is to use network technology to achieve equal positioning by both the borrower and lender in four aspects: information disclosure, risk appetite, term structure, and rights and obligations. In China, however, P2P was understood to be "person-to-person" by some market players. Their P2P platforms aimed to allow anyone with money to become a lender, and anyone who needs money to apply for a loan. Without proper setting, the online lending business was ripe for abuse.

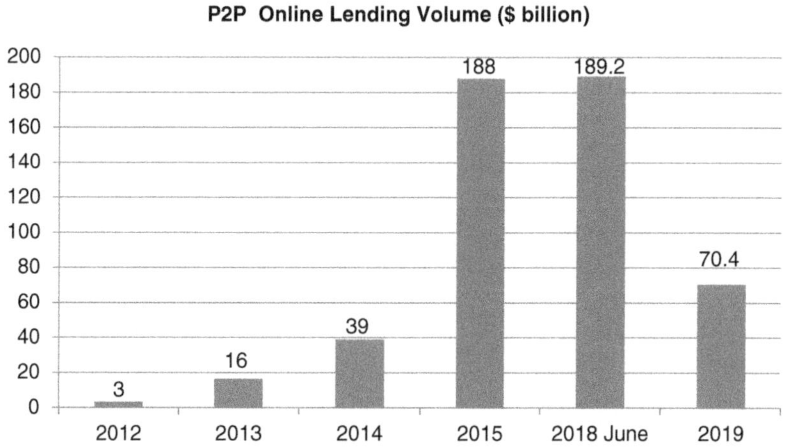

P2P Online Lending Volume ($ billion)

Figure 5.7   The abrupt expansion of P2P and its quick collapse (2012–2019)
Data Source: Wangdaizhijia.

There were numerous Ponzi scheme-like scandals where the lenders (before they fled and became unreachable) have pooled funds from investors without matching them to borrowers. Many promised returns three to four times higher than banks, but instead they wiped out the savings of millions of investors (some distraught victims even committed suicide), which sparked protests outside the headquarters of the banking and insurance regulators in Beijing. A government crackdown ensued, and the P2P lending industry has been in crisis mode ever since the regulatory scrutiny was tightened since 2017.

The increased regulation has seen total outstanding loans shrink to 491.6 billion yuan (US$70 billion) as of December 2019 from its peak of 1.32 trillion yuan (US$189 billion) in June in 2018, according to the industry data provider Wangdaizhijia. Meanwhile, just 343 P2P firms were still operating, down from 6000 at the sector's 2015 peak. To put the last nail to the scandal-plagued sector, in November 2019 the central government set up a task force, the China's Internet Financial Risk Special Rectification Work Leadership Team Office, to clean up risks in the online lending sector. It released a notice

Table 5.1   Online lending firms' dtocks dropping 90% from their highs

Company (Ticker)	When Listed	IPO Price (US$)	Historical High (US$)	Closing Price (US$)	Drop from High
HeXinDai (**HX**)	Nov. 2017 (NASDAQ)	10	17	0.39	97.7%
QuDian (**QD**)	Jan. 2017 (NYSE)	24	35.45	1.67	95.3%
XiaoYing Tech (**XYE**)	Sep. 2018 (NYSE)	9.5	20.3	0.95	95.3%
XinErFu (China Rapid Finance, **XRF**)	April 2017 (NYSE)	6	16.5	1	93.9%
YiRen Digital (**YRD**)	Dec. 2015 (NYSE)	10	55	3.76	93.2%
PinTec Tech (**PT**)	Jan. 2018 (NASDAQ)	11.88	15.1	1.26	91.7%
Weidaiwang (**WEI**)	Nov. 2018 (NYSE)	10	13.63	1.41	89.7%

Data Source: Public Markets, April 17, 2020.

and ordered all existing P2P platforms to exit the industry in two years.

The sector faces continued pressure in 2020 amid tightening regulation. Many smaller players have already shuttered their doors, and more platforms are likely to be shut down in 2020–2021. Even the most established platforms, such as those listed in the US stock exchanges, are not spared. By April 2020, every P2P internet finance company listed in the United States had seen its stock price dropped below its IPO price, and some of the stocks had dropped 90% below their historical high (see Table 5.1). (In July 2020, China police officially launched criminal investigation on Weidaiwang.)

The regulatory guidance from November 2019 suggested that P2P firms could transform themselves into qualified micro-loan lenders, if they could meet the requirement for commercial banks (e.g. registered capital of 1 billion yuan [US$143 million] and strict leverage limit). That's a promising path to survival for the (once) well-capitalized, listed lending firms as the demand for loans among small businesses and individuals is still growing in China. Their biggest challenge, as the next section shows, is that internet firms equipped with AI tech and Big Data are powerful players in the microloan market.

Micro Loans, Big Data Solutions

On January 4, 2015, the Internet firm Tencent launched China's first Internet-based bank WeBank (in Chinese infers a bank for "micro and many"). Among one of the five privately funded banks approved by China's central bank PBOC, the internet bank conducted all operations online, so that the borrower did not need to come to an offline outlet to receive the loan. Webank first used Tencent's own version of facial-recognition technology to verify the borrower's identity from a remote terminal; then, all the loan processing was completed online and paperless.

The ceremony was officiated by Premier Li Keqiang, and he pressed the "enter" button on a computer to send out the first loan of 35 000 RMB (US$5400) by WeBank to a truck driver. "One small step for WeBank is a giant leap for financial reform," said Premier Li. That statement was no exaggeration. Because the majority of the 1.4 billion population had no credit spending experience or even any interest in borrowing from financial institutions at all, there is no well-developed credit scoring system in China like FICO scores and other consumer credit record services in the United States.

Traditionally, most of the state-run banks in China favor lending to big institutions for two reasons. On the one hand, the major commercial banks have had long-term relationships with and knowledge of the big companies in China. The banks generally take an asset-based approach in credit assessment; hence, it is hard for individuals and SMEs to get the loans if a person or SME does not have collaterals or assets. On the other hand, historically the banks have limited data tools for credit risk evaluation on individual consumers and SME businesses, and the transaction costs for those small loans under the brick-and-mortar banking model is prohibitively high.

In other words, it is challenging for the commercial banks to acquire: (1) a large number of (2) small-scale borrowers (3) with good credit quality (4) at a low cost. As a result, while hundreds of millions of new middle class people are now financially active consumers, the existing financial system has not sufficiently covered the majority of the population for a long time. Many young, employed people do not have credit scores. That is an enormous vacuum that needs to be filled quickly as China moves into a consumption economy, and the internet-based Big Data approach may be the only efficient way to solve it.

By contrast, internet banks run branchless banking operations that can serve customers 24/7. Their cloud-based model requires much lower operational costs than the traditional brick-and-mortar banking model. And their parents' substantial databases of consumer behavior from internet businesses allow them to compile credit risk data in unconventional and innovative ways (before Big Data solutions were available, some P2P lending platforms probably got too innovative in risk management, **see the "Naked Loan and Flesh Payback" box**). As such, the internet banks can bridge a gap in the consumer economy, that is, providing small amounts credit to individual customers and SME businesses that cannot be reasonably priced within the formal banking system.

Naked Loan and Flesh Payback

In Shakespeare's *The Merchant of Venice*, the moneylender Shylock demands a pound of Antonio's flesh as collateral for a loan, and it must be taken from around his heart should he default. In modern-day China, the younger consumers such as college girls enjoy the status conferred by luxury brands, but they have no income nor credit card (in fact, most of the population has little credit history). Hence, flesh is back on the agenda for borrowers, thanks to a "legal" practice devised by shady lenders.

Some online lending platforms target college students buying laptops, smartphones, and other consumer electronics. The so-called "naked loan" (*luodai*) is a practice where young women can use their "body" instead of financial records as collateral. The term "naked" is a pun: for one thing, it means that no proof of financial capability is required when taking the loan, and therefore, such loans can be approved quickly after the "body" is submitted, often within minutes; for another, the nude photos (and videos, in the age of multi-media) of the female borrowers serve as the credit guarantee.

To get a naked loan, borrowers take naked selfies in which their ID cards have to be held in front of them. Both the front and back side of the ID card should be clearly readable. The borrowers also have to make a video (in which they also have to be naked), stating their name, the lender's name, the amount of the loan and interest, the date of payback. Because such loans carry high interest rates—typically a weekly rate of 30%—the young college girls often cannot repay the debt immediately. Then loaners will threaten to release the borrowers' naked pictures on the internet, or to expose their conduct to their parents and family, which often leads to sending more nude photos and videos as for more loans. Eventually, the lenders may propose a "flesh payback" (*rouchang*), for which borrowers will repay their loan with sexual services.

The scheme was uncovered by the public because 10 gigabytes of private information of naked loan users, including naked pictures and videos of female students between the ages of 17 and 23, were leaked from the online lending platform Jiedaibao. The public was shocked, and more so when it learned that the use of nude photos as collateral is not forbidden by law. Soon after that, Jiedaibao announced on its official Weibo account that it would take legal action to combat "naked loan" practices and that it would set up a one-million "anti-naked fund" to resolve the situation.

To evaluate individual consumers' credit score, Tencent is applying a social network-based methodology to the credit evaluation of internet users. On Tencent's social networking platforms like WeChat and QQ, as well as other Tencent-affiliated apps, more than a billion people have registered their personal data with effective identity certification. Again, take the first internet loan by WeBank, for example. For the loan to the randomly selected truck driver, Tencent's WeBank used internet-linked data mining tools to assess the credit background of the potential borrower. The driver was a club-member at the Tencent-invested logistics platform called

Huo-che-bang ("truck club"), which linked logistics operators that needed to ship cargo with truck driver companies.

At the time of the loan, this platform had one million drivers with 650 000 trucks as members, and it was serving more than 160 000 logistics company customers. For each club-member trucker, Tencent's platform had a large amount of information such as total distances travelled, total cargo volumes transported, the scope of orders handled, and so on. Some drivers had to prepay cargo freight, but commercial banks were not set to process such a small amount loan of that nature. WeBank, however, could resort to the data from the driver's operations on the mobile app, develop its own analytic model, and evaluate the potential borrower's creditworthiness.

Based on the driver's driving history on the mobile app and his online profiles on other e-platforms, WeBank's financial model gave the driver a high credit rating and granted a loan carrying a 7.5% interest rate, without requiring collateral and guarantee. In financial terms, WeBank's superior data sources and processing capability enables this internet bank to provide better priced credit to consumers (without sufficient information on the borrower, the borrowing cost on the loan would likely be higher).

Meanwhile, Alibaba has its own internet bank ("MyBank"), and before these official internet banks, the e-commerce giant (and its peers like JD.com) had already provided financing services for e-commerce merchants in the form of microloans or supply chain financing. For the web buyers, they provide personal credit lines that are essentially "virtual" credit cards. For example, Ant Group's Huabei ("just spend" in English) microlending serves individuals or companies that need short-term funding as they shop for purchases on Alibaba's e-commerce platforms.

According to data cited by the *Wall Street Journal* in December 2019, most users have credit lines of below 6000 RMB (US$865) and the average outstanding balance is about 1000 RMB (US$144). This credit line feature has helped

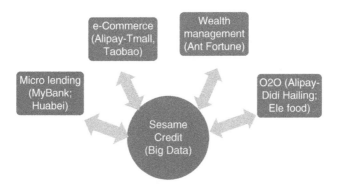

Figure 5.8   Two-way data Flow between Sesame Credit and digital businesses

Alibaba to boost its e-commerce transaction growth, as seen in earlier Singles' Day shopping festival discussion. At the same time, the firm's Big Data technology and dynamic credit risk management has helped keeping the delinquency of its consumer loans business at a lower rate than conventional banks and finance companies.

The linchpin is Sesame (*Zhima*) Credit, Ant Group's own credit-scoring system primarily based on online data (see Figure 5.8). While in the past, the Chinese were using Alipay simply to shop on Alibaba's Taobao e-marketplace, these days, the payment platform is so much ingrained in China that its hundreds of millions of users go to the platform to pay for nearly all household bills as well as O2O activities like ride-hailing and food order. Furthermore, its wealth management arm Ant Fortune also has hundreds of millions of users who manage their savings and investments there through Yu'e-bao and other financial products. In addition to shoppers' online e-commerce transaction records, Taobao and Tmall also have the history of dealings with tens of millions small businesses that sell or source goods on the two marketplaces.

As such, Sesame Credit has leveraged its giant user base—spanning about two-thirds of China's population—to cross-sell products and services. (Tencent has launched a similar credit score system based on WeChat pay transaction data.)

Besides facilitating payments, Ant sells mutual funds, makes short-term loans to individuals and small businesses, and offers insurance-like products. All these data on retail consumers enables Sesame Credit to work with those who may have little credit history at traditional credit agencies, which opens the door for Alibaba in two directions (see Figure 5.8).

On the one hand, Sesame Credit can facilitate more consumers to access new borrowing services such as home loans, car loans, and other types of installment credit. The platform algorithmically produces an individual (or small business)'s score and offers perks to users with good numbers (e.g. faster hotel check-in), while denying services such as train ticket purchases to those with bad scores. On the other hand, more credit transactions bring more data to the Sesame Credit database, which can serve as a foundation for Alibaba's new businesses lines (e.g. it has launched a cash loans product that is unrelated to purchases on e-commerce platforms). (The broad collection of personal data has led to data privacy issues, and Ant Group has noted that Sesame Credit is an opt-in feature for Alipay users, which is discussed in Chapter 8.)

When the coronavirus pandemic put China's economy into a pause in early 2020, the government encouraged lending to the smallest companies (a sector that contributes significantly to nationwide employment), for which the financial technology companies were part of the solution. As China lacks a standardized credit system, roughly 80% of the nearly 90 million small enterprises in China did not have a credit line with a bank, according to Tencent-backed WeBank (2019 media report).

Furthermore, the size of small enterprises' loans is also a mismatch for traditional banks. The banks rarely issued loans of less than 1 million RMB, while the average size of a small business' borrow on Ant-backed MYbank was about 9900 RMB (US$1,469), according to Ant Group (2019 media report). What the battered economy needed was smaller loans and quicker processing than before. The digital financiers delivered the solution: their systems could quickly approve a borrower's credit limit according to its online behavioral data in just a few minutes, closing the finance gap for the small enterprises in China.

AI and Alt-data for Wealth Management

For most Chinese, their first experience with personal wealth management came after they have had a smartphone. Their investment activities started modest with their spare changes, which were already left in the mobile payment apps after their frequent e-commerce transactions. In fact, the very first online investment product, offered by Ant Group, is called *Yu'e Bao* (literally means "leftover treasure"). It is the best example of financial innovation in the mobile economy, and its decreasing relevance of late illustrates the market shift into more sophisticated, data-driven investments.

In June 2013, the first Internet Money-Market Fund (I-MMF) product Yu'e-bao was launched, which was offered to internet users as an alternative to traditional short-term deposit at commercial banks. The explosive growth of this simple internet product shocked traditional financial institutions. Within six months, the Yu'e-bao product made the underlying manager, Tianhong Asset Management, the first investment fund in China market history to exceed AUM (assets-under-management) 100 billion RMB. In less than a year, Yu'e-bao's AUM exceeded 600 billion RMB (more than US$90 billion). At its peak in 2017, its AUM reached an astounding 1.7 trillion yuan (approximately US$250 billion), standing as the largest money market fund in the world (see Figure 5.9).

The exceptional popularity of the I-MMF products is the result of internet firms identifying and solving several structural issues of the financial market in China with the internet tools. **First**, Chinese households have one of the highest savings rate in the world, so they have trillions of cash in bank deposits. They are not happy with the low returns from the deposits, but they are not necessarily comfortable with the

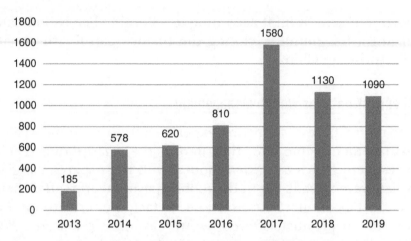

Figure 5.9   The rise (and recent fall) of Yu'e Bao (billion RMB, 2013–2018)
Data Source: Tianhong Asset Management (year-end AUM).

risks of public stocks or real estate. Therefore, Chinese investors are constantly in search of risk-controlled alternatives to demand deposits.

Second, thanks to the low operating cost of mobile internet, Yu'e-bao has no investment threshold (which is required by most wealth management products sold at banks), and its minimum initial investment is as low as 1 RMB (15 cents in US dollars). Meanwhile, it accumulates large numbers of "leftover capital" in the society via the internet (in the same way as the celebrity online novelists are supported by a large number of *diao-si* readers who pay a miniscule amount to read an installment). By pulling a big pool of money together, Yu'e-bao's manager Tianhong could negotiate higher deposit rates (for the benefit of its users) when it put the capital into the interbank market.

Third, the I-MMF products have high liquidity because of simple online procedures, and they can be easily processed on smartphones and other gadgets. Those I-MMF products like Yu'e-bao can generally be redeemed on the same day and in many cases within the same hour. By contrast, traditional money-market funds (MMF) require two to three days to complete the redemption process. In other words, I-MMF products like Yu'e-bao have the high-yield benefit of a MMF, but maintain the same liquidity as a bank deposit. During its peak year of 2017, Yu'e-bao fund processed 13.47 trillion RMB investments (about US$2 trillion) and 12.7 trillion RMB redemption.

From a different perspective, the essence of the Yu'e-bao success is simply using the internet as a better information channel and a more convenient transaction platform. As such, its investment return has mostly followed the corresponding market return. In recent years, the financial markets have entered into a new era of "low yield" and the return on Yu'e-bao has also dropped, leading to the decrease of its AUM (see Figure 5.9). In early 2020, the Yu'e-bao's return dropped below 2% for the first time since its 2013 inception, and it saw more users pulling money out to seek higher returns elsewhere.

Overall, the Yu'e-bao case illustrates future directions of the increasingly digitalized investment markets. First, digital technology will change the investment product distribution landscape in China. As the following examples show, internet platforms, robo-advisors, and voice-enabled technologies are the channels of the future to reach clients. Second, in today's low-yield environment, artificial intelligence and Big Data technologies may hold the key to the future of active management, where fund managers could potentially generate "alpha" (industry terminology for fund outperformance) and deliver high returns to their investors.

Digital Channels. China's emerging middle class (estimated by McKinsey to reach 400 million by 2020) presents a tremendous opportunity for the asset management industry. As individual investors become more sophisticated in wealth management, they are making a gradual shift from bank savings (Yu'e-bao and equivalent products from internet firms) or fixed-income products to more-diversified wealth-management products such as stock funds and VC/private-equity types of alternative investment products. China's conventional financial institutions (primarily focusing on high-net-worth individuals) simply don't have enough financial advisors to meet the need.

Because Chinese customers are relatively immature in terms of their investment experience, the requirements for personal guidance and education-oriented services are higher, which is another operational challenge to the wealth-management firms. As such, they are implementing AI technologies to make tailored products and services more accessible to ordinary users.

For example, in 2017, Ant Fortune, Alipay's wealth management platform, launched *Caifuhao*, a corporate account offering on the Ant Fortune platform that gives financial institutions access to a suite of AI capabilities and other services, including better user connectivity, user profiling, operational optimization, and smart marketing. Two years later, in June 2019, Ant Fortune announced that 80 of all 124 asset management firms in China, or six out of ten, had been using the company's AI-powered corporate Caifuhao accounts to reach more customers with better customized offers, at a lower cost.

Another tech solution for the lack of financial advisors is robo-advisors. For example, the machine-based financial advisor **Licaimofang** considers itself as China's first "All AI" robo-advisor institution, which provides automated financial-advisory service based on a totality of ordinary customer data. By early 2020, the firm's AI-driven decision-making process had transacted more than 10 billion RMB (about US\$1.5 billion) for its clients. For China's digital middle class, it is more likely than its Western markets peers to accept robo-advisor services to start its journey of financial investments

Alpha (α) Generation. In the current low-yield environment, fund managers across the globe are using AI and Big Data technologies to try making intelligent investments— better informed buy-and-sell market decisions—and bring in higher returns above general market return (alpha). In the developed markets of the United States and Europe, alpha seeking is increasingly difficult. China's market, however, is much younger (the stock market only started since the early 1990s; see the author's earlier book *Investing in China—Opportunities in the Transforming Stock Market* for reference) and less efficient, such as that active management and alpha generation remains possible, especially with the help of new technologies.

To make active strategy traders happy, the speculation level in China's stock market is extremely high. The most commonly used indicator for market speculation is the "average stock turnover rate", which is defined as:

$$Average\ stock\ turnover\ rate = \frac{Total\ annual\ trading\ value}{Annual\ average\ market\ capitalisation}.$$

A higher turnover rate indicates more frequent trading or a shorter average holding period for stocks, and is thus an indication of prevalent speculation. According to a 2002 Dow Jones Indexes report, the annual average turnover for China's A-share market during the 1994–2001 period was more than 500%. This can be interpreted as each stock on average being traded five times in a year, or in other words, the average holding period for a single stock was less than two months (one-fifth of a year). By contrast, in developed markets the typical holding period is about two years. (See the author's earlier book *Investing in China—Opportunities in the Transforming Stock Market.*)

The speculation level in China's stock market thus is extraordinary, which in turn, leads to the high volatility of the stock prices. According to Bloomberg's news report in 2019, China's main stock index has the highest volatility among global peers (see Figure 5.10).

In a market characterized by such dynamic and volatile trading, one would naturally expect that AI algorithms will emerge to emulate successful stock pickers. **Zheshang Fund Management Co.**, which manages US$6.5 billion in 2019, has made significant progress on that. In the same way as AlphaGo—which learned about the *Go* chess from the historical games played by best human players, the Zheshang-developed AI machines first

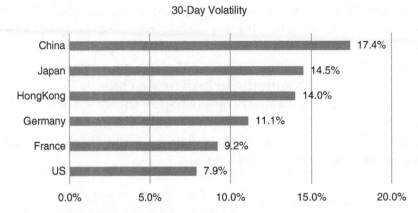

Figure 5.10   The high volatility of China's stock market
Source: Bloomberg, July 2019.

analyzed the strategies of more than 80 of the nation's best-performing fund managers, drawing from public disclosures of their holdings. They were also fed with stock recommendations from 500 star analysts and industry experts for "training".

Unlike one single AlphaGo, the Zheshang AI fund has developed more than 300 machines based on different investment logic and methodology, which are referred to as its "robot managers". They analyze more than 3000 listed Chinese stocks and develop hundreds of investment decisions, which are put to compete to find the best solution. New robot managers are constantly added with new learnings and new strategies, and worst-performing ones are removed. After about a year in training, Zheshang in late 2019 launched the nation's first AI-based fund run on recommendations from its machines. According to media interviews, the investment performance by the AI managers was mediocre at the beginning, but their learning and improvements has accelerated with cumulative trainings—and investment returns kept climbing higher.

The biggest challenge, according to the fund manager, is that the machines haven't had sufficient time and data to learn, especially for newly listed companies and new industry sectors in China. Furthermore, to make the best use of the patterns the robot managers identify from market data, the fund integrates machines with skilled professionals to optimize investment decisions, which is referred to as "AI + HI (human intelligence)". (Interesting to note that, the fund initially also considered the AlphaGo Zero approach—providing investment rules for the machines and have them self-training. While AlphaGo Zero easily beats AlphaGo in *Go* chess (see Chapter 1 for details; the Zheshang team found that approach didn't work well in investments. The author speculates that it may have something to do with unexpected human interventions in the stock market, such as Chinese governments' policy changes and Twitter tweets by Donald Trump).

Alternative Data

Thanks to digital technologies, traders today can look into data sources beyond the typical company filings, earnings calls, or fundamental datasets. As such, they are referred to as "alternative data" (or alt-data). Early alt-data sources consisted of credit card transactions, web scraped data, satellite imaginary, weather forecasts, employee sentiment analysis, and geolocation data from cell phones.

The advent of the smartphone a decade ago brought more data possibilities into the mix. For example, it is a source of geolocation data linked to user's access to specific mobile applications. Also, its mobile payment apps have the complete records of users' e-commerce transactions. Furthermore, it has the users' social-media data, providing insights on the virtual community's personal interests and social trends. (Some of these sources are controversial when it comes to consumer privacy issues, see the detailed discussion in Chapter 8.)

Based on alternative data sources, Big Data analytics and machine learning techniques enable investors to explore new patterns across the global capital markets. For example, they can help analysts to generate correlations and trends unidentifiable by the human brain. Also, natural-language processing (NLP) and sentiment analysis can enable traders to dissect investor call transcripts, seeking out verbal signals that point to changing conditions. In another example, investment firms may use the satellite image to track environmental polluters, such as coal burning without much treatment, and to make a prediction on their ESG compliance.

Another good example of "AI + HI" in China's stock market is the partnership between **Microsoft** and **China Asset Management Co.**, Ltd. (AMC), one of the largest investment houses in China with US$153 billion in assets under management, for data-based quantitative investments. It has been a while for traders to use managers analyze big data to make investment decisions, however the tricky part is collecting the right data from the mountains of information available, because financial data is notoriously "noisy", meaning there are many potentially misleading signals that need to be filtered out. Right now, even the best quantitative traders may only use a fraction of all the data out there, meanwhile digital technologies are still turning more non-traditional data into the mix (see the **box "Alternative Data"**).

In September 2019, AMC and Microsoft announced that their researchers had developed a new AI model that sifts and analyzes vast amounts of real-time financial data. With machine learning (ML)—a subset of AI, they created two sophisticated methods to predict the market. The first is called a "spatiotemporal convolutional neural network", which identifies patterns in volume-and-price data by estimating what is going on among different variables and isolating important factors to watch. The second is called a "time-varying attention model", which compares those factors with what is happening in the market at any given moment. It then combines them in a time-varying way to identify dynamic market trend patterns. These analyses are provided to investment professionals to better their investment decisions.

In both examples, AI and Big Data's key advantage comes from their ability to perform deeper analysis on a wider set of data than what would be humanly possible. Machine learning (ML) augments the quantitative work already done by security analysts in three ways (see Figure 5.11). Therefore, the digital giants are evolving into asset management powerhouses themselves, as their digital expertise and direct-to-consumer marketing prowess far exceed even the biggest and best-resourced

Figure 5.11   ML augments quantitative investments

asset managers. (E.g. Alibaba could use consumer sentiment data captured by its multiple e-commerce marketplaces to predict retail sector stock performance with a heightened level of insight.)

Blockchain and SCF: Alliance of Alibaba and ICBC

"Blockchain is not a bubble, but Bitcoin might be. Bitcoin is only a tiny application of blockchain", remarked Jack Ma, the founder and former Chairman of Alibaba, during the 2018 World Intelligence Congress, "At least inside Alibaba, blockchain must be a solution that addresses privacy and security issues in the digital era". Ant Group clearly follows that logic. Since it started the blockchain business, it has focused on building up an open and inclusive enterprise-level blockchain ecosystem, while developing real-world applications for various industries.

The same is true for its fellow internet firms and traditional financial institutions. After the initial Bitcoin and then Ethereum (smart contract)–based tokens, the blockchain technology in China has reached its third phase, where the major players focus on enterprise blockchain solutions (see Figure 5.12).

In April 2020, China's National Internet Finance Association of China (**NIFA**) published a report about financial blockchain projects in China. Based on statistics the Cyberspace Administration of China (**CAC**, under whose regulations enterprises providing blockchain-based services must register as such with regulators and must collect real name and identity

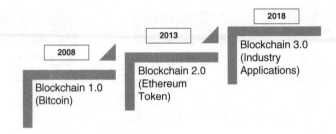

Figure 5.12   Blockchain from 1.0 to 3.0

of users of the blockchain service), the NIFA report found that financial institutions seem to be the most active in the field, as 40% of the organizations involved in blockchain tech in China are "fintech service providers, banks, fund companies, insurance companies, microfinance companies, commercial factoring companies, and so on".

The NIFA report also found that blockchain-based supply chain financing (**SCF**) is where the financial institutions are focusing on. According to its survey, 32.6% of all organizations were said to be working on supply chain financing, making up for a major part of blockchain implementation (see Figure 5.13). Trade finance and insurtech account for 11.2% of the market size each, while cross-border payments and asset securitization further make up 7.9% and 6.7%.

Indeed, blockchain has been applied into the supply chain financing (SCF) model in China, and the benefits are already

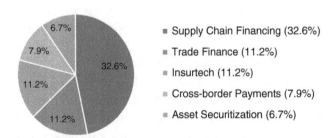

Figure 5.13   Supply chain financing the leading blockchain application

evident, especially for SMEs (small- and medium-sized enterprises). The main form of SCF is "factoring financing", where a business uses its accounts receivable (i.e. invoices) to receive financing for its immediate cash needs. In a supply chain, the so-called "core enterprise" (e.g. a major carmaker), which acquires components from numerous suppliers, has a long-term, well-established relationship with commercial banks. As such, the core enterprise itself enjoys favorable credit relationship with the banks and easily draw banking loans at an attractive interest rate.

On the other hand, the "good credit standing" of the core enterprise (also known as the "key company" of the supply chain) is not easily transferable to its suppliers. Because the transaction flow and cash flow information among the multiple layers of suppliers is separate and opaque, the commercial banks have no effective risk control method to offer financing, in favorable terms based on the good credit standing of the core enterprise, to suppliers not immediately dealing with the core enterprise, especially the small- and micro-sized enterprises at the lower end of the industry chain. The small players cannot find attractive financing, whereas the banks have to pass the small loan businesses that are indirectly backed by the core enterprise of substantial scale. In practice, banks often only offer factoring financing for the first-tier suppliers of core enterprises.

Thus, the blockchain technology is used to solve the data and trust issue in SCF, so that suppliers who are tier 2 or even further up the supply chain can apply for factor financing. Blockchain is the ideal technology to ensure that all data of the cash flow and trade flow among core enterprises and all-tier supplier (their performance and related risk factors) can be shared in an authenticated manner with the banks (and themselves), even when there is no direct relationship between them. Thanks to the immutability and traceability nature of blockchain, the data exchange is secure, verified, and trustable throughout the blockchain in real time, so that all members

Table 5.2   Chinese banks/internet firms launch blockchain-based "real economy" applications

Banks/ Internet Firms	Platform Infrastructure	Supply Chain Finance	Trade Finance	ABS Asset Backed Securities	Digital Certificate	House Rental
ICBC	√	√	√	√	√	
ABC		√				
BOC				√		
CCB		√				√
CMB			√	√		
Ping An Bank	√	√	√			
Ant Group	√	√			√	
WeBank	√	√			√	
JD Digital	√	√		√	√	

Data Source: ICBC 2020 Blockchain White Paper (ICBC—Industrial and Commercial Bank of China; ABC—Agriculture Bank of China; BOC—Bank of China; CCB—China Construction Bank; CMB—China Merchant Bank).

of the supply network can access the information anytime. As such, the digital credit credential of receivables of suppliers are formed on the blockchain, based on which banks like ICBC could extend credit to the upstream and downstream companies in a supply chain.

Just like Ant Group, WeBank and other internet firms, ICBC and its banking peers (see Table 5.2) also provide blockchain platform infrastructures for the market (**BaaS**—blockchain as a service). The BaaS services from these giants are the links between blockchain architecture and the enterprise-level blockchain projects. They enable blockchain entrepreneurs to focus on building new blockchain applications, leading to a wave of blockchain startups in China.

Going forward, the market would see more convergence between the digital tech companies and traditional financial institutions in China. The major commercial banks have already fully developed internet channels and platforms for their customers. They have stepped up technology investments as part of its push to become a next generation "smart banks". About three quarters of commercial banks are planning to, or have already, put in place strategies for digital transformation, according to an October 2019 survey by the National Internet Finance Association (NIFA) of China.

Naturally, they are developing partnership to find synergies. In December 2019, the ICBC, the world's largest bank with over US$4 trillion assets, announced its partnership with Alibaba and Ant Group to enhance its banking service. The companies said they would explore new services on a broader scale beyond payments and e-commerce, including areas such as blockchain-powered cross-border remittance and cloud-based applications. More recently, China Everbright Bank developed its supply chain finance product on Ant Group's BaaS platform to expedite financing related to rework (from the coronavirus pandemic) funds. According to Alibaba's press release, the company has partnered with over 200 banks, including Huaxia Bank, Shanghai Pudong Development Bank, and China Citic Bank.

In summary, China is the world's largest fintech market, and it is a global leader in almost every aspect of fintech developments. The alliances between the tech firms and financial institutions will further drive innovations in the monetary and financial systems. China's fintech success, however, is not just from unprecedented technological innovation, but also from integrating tech tools with real economy needs, which is best illustrated by its distinctive approach to Bitcoin and blockchain technology.

CHAPTER 6

O2O and the Shared Economy

- Sperm Donation, O2O, and the Sharing Economy
- Didi Taking over Uber in China
- Mobike and Ofo: Bike Sharing of Yesterday
- Umbrellas, Sex Dolls, and Everything to Share
- Meituan, Didi, and Ele.me: All-in-one On-Demand Services Platform
- From MaaS to OaaS—a Smart Pie

Sperm Donation, O2O, and the Sharing Economy

In 2015, Alibaba's e-commerce platform Taobao and KingMed Diagnostics, China's biggest third-party medical laboratory group, jointly offered great deals online for sperm and donation paternity tests. Within the three days of the campaign, the sperm donors were able to register their personal information online, avoiding an embarrassing first visit at the sperm collection centers. Volunteers only needed to provide their name, last six digits of their ID card, and email address to complete the online registration. After their sperm donation, each volunteer could receive a subsidy somewhere between approximately US$475–$800.

Within 72 hours, more than 22 000 men reportedly signed up for sperm donations at one of seven sperm banks nationwide. The Taobao campaign also offered paternity testing services without customers having to initiate a hospital visit costing just above US$100, a huge discount to the average cost of about US$650. A testing kit was sent to the customer after he or she placed an online order. The consumer only needed to collect and send a saliva sample to a test center, which would provide the test result after 10 working days. During the same period, Taobao attracted 137 applications for paternity tests and 4060 for sperm fertility testing.

The development of mobile fintech and location-based data technology are driving the growth of online-to-offline (O2O) transactions like the "sperm donation" example. O2O can be broadly defined as the integration of offline business opportunities with the activities on the internet. In its most popular form, O2O means attracting retail customers online, and then directing them to offline transactions, hailing taxis, dining out, seeing movies, and so on (see Figure 6.1).

In addition to business model transformation (e.g. China's movie industry is a perfect example of the O2O revolution; see the related discussion in Chapter 3), the O2O boom has also stimulated "shared economy" applications in many economy sectors. For instance, the O2O business for medical and

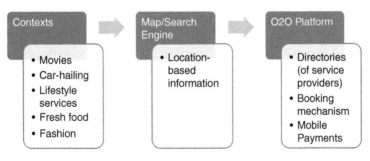

Figure 6.1   Location/map searches are the key to an O2O business model

domestic services started to take off in response to the strong demand from an aging population in China. Healthcare providers are implementing remote patient monitoring to stretch their footprints to underserved patient populations while substantially saving costs for patients with chronic diseases.

As the growing Chinese economy brings more people access to personal wealth for the first time, the sharing economy has profound implications and vast potential. Currently, China has more than 300 million middle class people, equivalent to the entire population of the United States, and that number is expected to double in just a few years. Even at the 600-million level, however, that middle class is still only a minority of the total 1.4 billion population. More strain will be put onto the cities, infrastructures, and environment when hundreds of millions of more people ascend into individual prosperity.

For example, the year-on-year growth rate of auto sales had 30% and 50% annual growth in 2010 and 2009, but since 2011, its growth has to be slowed down by city limitations on new car license plates because the industry has turned into a heavy user of resources and a major contributor to pollution. As such, the mobile apps offered by the US car-service and pooling company Uber and its homegrown rival Didi have been widely embraced by Chinese urban residents. (Didi achieved its dominant market position after a US$35 billion merger with Uber China in 2016. See the detailed case study in the following section.)

What's interesting is the fact that before it merged with Didi, Uber's service had taken off in China much faster than it did in the United States. For example, Uber found that even when it came to the initial customers in a region—the first one million users in a city, its growth in China was much faster than what it had seen elsewhere. Uber's top three most popular cities worldwide—Guangzhou, Hangzhou, and Chengdu—were all in China.

The main reason for that, of course, is the large urban population and the high concentration of city residents following the central government's urbanization push. There are about

200 cities in China with more than a million people. The Chinese urban residents have generated strong, new demand for goods and services, and the cities need to adopt "shared economy" models for their scale, density, and necessity. Another reason is that Chinese customers are ready adopters of social network and mobile payment technologies. Almost all urban residents have installed e-payment apps on their smartphones and linked to their bank accounts accordingly, so that they can complete mobile payments directly after taxi rides.

Therefore, China is perhaps the best lab and market for "shared economy" applications. Many industries in China, just like the auto industry, will soon be caught in a dilemma in the wake of energy supply strains, traffic woes, and a deteriorating environment on a broad scale. Hence, the rise of "sharing economy" that emphasizes sharing over ownership. The mobile internet, smartphones, and payment systems have created a new social coordinating mechanism, and what the market sees is a seemingly endless potential to put goods and labor that are less than fully utilized to productive use, as shown by the case studies in this chapter.

What's also interesting is that car ownership in China as a status symbol is fading in parallel to the decline of registration of new individual cars. According to Bain & Company's market surveys, in 2014, almost 60% of the consumers in tier 1, 2, and 3 cities believed that owning a car improves one's social status. However, in three short years, car ownership has lost some of its luster. In 2017, less than 50% agreed (see Figure 6.2).

This is an encouraging sign that the young generation of Chinese consumers is defining the "China Dream" with a new context of personal prosperity. They are acutely aware of the resource constraints and income disparity arising from the economic boom, and many of them have actively accepted the sharing model of consumption. Equipped with digital technologies, they are creating a more inclusive and sustainable model of economic growth, because IT innovations intrinsically emphasize sharing over ownership.

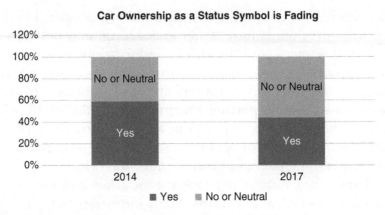

Figure 6.2　Does owning a car improve one's social status?
Data Source: Bain & Company.

While both China and the world are at the beginning of a new lifestyle that moves away from ownership into sharing, such a shift probably is the most important in China than anywhere else, due to its sheer scale of new customers. The new China Dream *has to be* different from the traditional "American Dream"—a big car, a big house, a big dog, and Big Macs for all. Otherwise, the Chinese consumers need to put their heads together to find another planet. (Notably, everything in the "American Dream" today is even bigger than 20 years ago, when this book's author first arrived in the United States for law school graduate studies.) As such, the shared economy in China will have profound implications for the increasingly stretched world.

Didi Taking over Uber in China

As in many foreign cities, demand for cross-town transportation is at the heart of an urban lifestyle of modern China. Although more individuals are buying cars, the demand for taxi service is steadily increasing as the population in many cities has exploded in recent years due to urbanization developments. However, Chinese cities limit the base fare that taxis can charge customers. In places where the government-mandated taxi fares have been kept low or even cut, taxi drivers enter into a strike from time to time, making hailing a taxi next to impossible.

The car-hire apps emerged in 2012 to allow smartphone users to book and pay for taxi rides or limousine services using mobile apps, and it has changed the way many Chinese travel around their cities. The two leading car-hire companies, which started their businesses in taxi-hailing before expanding into carpooling and premium car services, were backed by two of China's biggest internet companies, Alibaba and Tencent, respectively. Before their merger, Alibaba invested in Kuaidi (meaning "finding a taxi swiftly") and Tencent invested in Didi (meaning "honk honk" of a taxi).

Uber, the San Francisco-based car-sharing app company, was relatively a latecomer to the ride-hailing service market in China. Uber began tests in China in late 2013 in the southern cities of Guangzhou and Shenzhen, offering a service in which customers could hail rides from licensed limousine companies. Its official full-scale expansion in China did not occur until December 2014, when Uber China sold an equity stake to China's search engine giant Baidu and integrated its service with Baidu's map application and payment system. Consequently, the three companies—each backed by one of the BAT—competed fiercely in the China market (see Figure 6.3).

After their launch in China, the Didi and Kuaidi apps caught on instantly with China's urban residents. The congested large

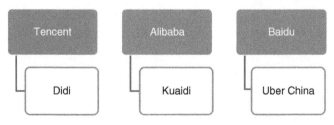

Figure 6.3   The competitive landscape before Didi, Kuaidi, and Uber China's merger

cities, the insufficient public transit networks, and the relatively large fleets of inexpensive taxicabs make the Chinese market perfectly made for taxi-hailing mobile services. But trying to get taxi drivers in China to use cab-hailing application was not easy. In the earlier days, both companies had to find creative ways to attract, train, and convince taxi drivers who were not tech-savvy. They competed head-to-head on the streets to promote their respective app and corresponding mobile payment systems (Alibaba's Alipay and Tencent's WeChat Payment).

After a price war reportedly involving more than US$300 million in total rebates for taxi drivers and customers, the two companies did entice many cab drivers to sign up with Didi and Kuaidi apps. (In fact, almost all taxi drivers in big cities now have multiple smartphones and tablets installed in their cabs for various apps.) However, the cumulated subsidies were too costly for the internet firms. In February 2015, those two companies decided to enter into a US$6 billion merger to end the price war (possibly a symbol of "marriage" and "good will", the merger news was announced by the founders of both sides on Valentine's Day).

Patients, Nurses, and Safe Shots

In 2015, Uber paid out a large amount of bonuses to Chinese drivers to build its brand awareness and market share. An unintended consequence was that groups of fraudsters developed specialized software to get paid on fake rides. As such, Uber's performance of one million rides a day was put to question.

For example, one driver might have received a job at the airport, but he still wanted to get paid for the trip there while the car was empty. He could go into one of several invitation-only online forums and request a fake fare from fellow fraudsters. The drivers refer to themselves as "patients", and those professional ride-bookers were referred to as "nurses" because their task was to use specialized software to put an "injection", a location-specific ride request (from no real customer) near the driver.

After the "injection", the "patient" (driver) headed to the location selected remotely by the "nurse" (booker), and then completed the trip while the booker monitored remotely to ensure the fake journey showing up on Uber's GPS tracking software. After all was said and done, the driver could collect the fare and a bonus up to three times the fare, while the "nurse" got paid by the driver for the fare reimbursement and a small fee.

Alternatively, the driver could do it him- or herself. For that, the driver needed to invest in a modified smartphone that could operate with multiple phone numbers and multiple corresponding Uber accounts. The driver could use one number to act as a rider, request a lift, and then accept it from his or her driver account. Such Uber driver and rider accounts were also for sale, according to advertisements on e-commerce marketplaces, which offerings also promised to be able circumvent Uber's background checks.

On those online marketplaces, the listed offerings even included modified smartphones that could be used to pretend to be a new user each time they placed an order. It could show unique 15-digit identity numbers to make Uber's software believe they were different phones every time the modified phone was used. With this feature, the users could "permanently" enjoy the "first ride" promotion (an additional credit for first-time users). Of course, the companies worked hard to crack down on these scams, and a software upgrade in late 2015 apparently closed out the loophole.

The combined company, Didi Kuaidi, had a near monopoly on mobile hailing traditional taxis, and it also controlled more than 80% of China's private-car-hailing market, according to research firms' estimates at the merger. Then it continued the fierce subsidy battle with the remaining Uber China, which, as a foreign company in China, was struggling with local issues in the complex and controversial car-hiring business environment. For example, because Uber offers large amount of cash as bonuses to Chinese drivers to get them on board, some drivers game the system by playing tricks on rides using specialized software. (**See the "Patients, Nurses, and Safe Shots" box**.)

In August 2016, after some US\$2 billion of losses (and potentially more), Uber China was bought out by Didi in a US\$35 billion deal and (former) Uber CEO Travis Kalanick became a board observer at the merged company called Didi Chuxing ("honk honk going out"). The price war was over, but all the subsidies and bonuses had educated drivers, passengers, and the public of smartphone car hailing, mobile payments, and the concept of "shared economy".

According to Bain & Company's market surveys, Chinese consumers have brought mobile car hailing and bike sharing (see the more detailed discussion in the next section) into the mainstream, with 62% and 73% respondents used these two mobility solutions, respectively, in 2017. By contrast, there were less than one-third of respondents who used mobile e-hailing in Germany (29%) and the United States (23%), and further less used bike sharing (see Figure 6.4). As is seen in the following sections, after the "education", the Chinese users fully prepared themselves for bike sharing and virtually everything sharing based on smartphones and mobile payments.

After vanquishing Uber in China, Didi has become the default app for Chinese people moving around and the world's largest ride-sharing company by number of rides. In a 2017 funding round, the ride-hailing startup was valued at US$56 billion. Didi received US$10.9 billion in cumulative investments from SoftBank—the same leading investor in Uber, according to a Bloomberg news story in January 2020. In Uber's 2019 annual report released in early 2020, Uber estimated its roughly 15% stake in Didi to be worth US$7.95 billion, implying a valuation for the private Chinese company of about US$53 billion.

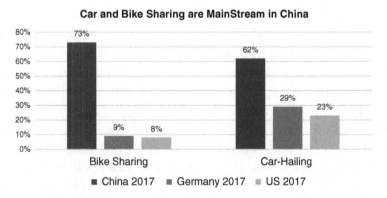

Figure 6.4   Have you used the mobility solution that you are familiar with?
Data Source: Bain & Company.

Valued at more than US$50 billion, Didi is the "D" in "TMD", the acronym for the three new tech leaders after BAT (see the related discussion in Chapter 2). Didi has also started overseas expansion, including in January 2018 when it acquired control of 99, a Brazilian ride-hailing company. It set a goal to turn profitable in 2018 and the company was reportedly on course for a public listing as early as 2018.

Then Didi was confronted with its biggest crisis ever since it was founded six years ago. In mid-2018, two young girls, just three months apart, were raped and killed separately by the drivers on Didi's popular hitch-hiking service. The incidents shocked China and the increased attention from regulators and police has made it harder for the company to grow. (E.g. government regulation of the industry has tightened curbs on unlicensed drivers. See the detailed discussion relating to user privacy and data regulation in Chapter 8.)

Meanwhile, new rivals such as CaoCao, Dida, and Shouqi Limousine & Chauffeur cropped up. Meituan Dianping, the life service platform and the "M" in "TMD", also launched its ride-hailing projects. Therefore, the company is stuck in the price war—with new players. Despite its expansion in overseas markets, which contributed little to the group's revenue, getting to profitability in the main Chinese market remained a struggle for Didi. When its US counterparts Uber and Lyft went public—and their shares tanked—many investors began to question whether Didi could ever turn profitable. (Just like Didi, Uber had significant losses from its operation. Uber IPO-ed in 2018 at a valuation of over US$82 billion. By early 2020, Uber's value had dropped below US$45 billion.)

In 2020, the coronavirus outbreak dealt the company a particularly harsh blow. Drivers were reluctant to put in hours, and passengers cut back on trips because both were disinclined to travel with other people in a small, contained space. As a result, Didi's passenger count tumbled, compounded by rising operating costs, such as distributing one mask per day to drivers and installing plastic sheets in vehicles to separate drivers from their passengers. Even though the Chinese economy began to reopen in the second quarter of 2020, the virus set yet another obstacle on Chinese ride-hailing app's long road to profitability and derailed what was already a fragile recovery from the 2018 crisis.

Nevertheless, Didi is still a powerful platform in an enormous consumer market. Didi retains a dominant position in Chinese ride hailing, with 93% of total daily active users in 2019, according to the analysis of Bernstein, and it has also expanded into the bike-sharing market to expand its mobility system. In July 2019, Toyota Motor Corp. invested US$600 million into Didi's self-driving business and car-leasing unit Xiaoju Automobile Solutions, which was viewed as a vote of confidence by the media. (Toyota recently announced that it was transforming itself into a mobility company, whose competitor is Apple instead of Volkswagen.)

Based on its larger user base and mobility system, Didi is expanding into more new businesses for growth and profit. For example, during the coronavirus pandemic, Didi launched delivery services in major Chinese cities. The competition in food and delivery services, however, is no less fierce than the ride-hailing market. The largest existing player is none other than Meituan Dianping, which has consolidated the food delivery market after a similar price war and also moved into the ride-hailing and bike-sharing businesses. After the following sections on bike sharing and virtually everything sharing, we look into the all-out war between Didi and Meituan in a much bigger "sharing economy" context.

Mobike and Ofo: Bike Sharing of Yesterday

In the 1980s, China was known as the "Bicycle Kingdom", with freeway-sized bike lanes accommodating more than half a billion commuters during rush hours. As a result of China's reform and opening-up policy in the 1990s, transportation means were upgraded accordingly. Bus routes and subway systems underwent massive expansion; electric scooters became popular; taxis became affordable; and in more recent years, individual families began to own their own cars. Car lanes expanded while bike lanes shrank.

It is therefore fascinating to see the humble bicycle making an unexpected comeback in China's cosmopolitan cities, thanks to the digital revolution. Since 2016, the sidewalks of all major cities are filled with a rainbow assortment of colorful sharable bicycles—yellow, orange, green, or some other color combination. The specially designed bikes are equipped with GPS and proprietary smart-lock technology, and users can easily use mobile apps to find a bike near them, reserve and unlock it, then complete their trip by simply closing the lock anywhere regular bike parking is allowed.

Because they are practical, environmentally friendly, convenient, and cheap, bicycles have become the No. 1 transportation choice for urban residents, students, and young professionals in China's leading cities. Whereas many cities also have their own bike-sharing setup, there are three special characteristics of Chinese bike sharing that make it particularly popular.

First, it is the most practical way of traveling the last mile. Even with more roads, buses, subways, and taxis than ever, urban traffic is still challenging, especially for short commutes from people's homes to bus and subway stations. In those cases, the destination is too far away to walk, but not so far as to call a taxi and then have to wait for it. According to China's State Information Center, using a shared bike + subway combination is 17.9% more efficient than a private car for city travel.

Second, the bicycles are dockless so users don't have to worry about returning them to a specific place after a trip. They can be used and left anywhere, anytime without the confinement of a docking station (other people then locate the bikes using GPS), creating a better user experience as well as higher profit margins. In contrast, other bicycle-sharing systems require users to pick up bicycles and return them to a fixed dock, such as Citi Bike in New York City.

Third, the price per ride is cheap and conveniently paid. Users can ride a bike over long distances at incredibly low prices (usually 1 RMB/$US.15 for 30 minutes, or even less). All it takes is a smartphone to find the nearest bike via GPS, scan its QR code, pay the minuscule fee via WeChat or Alipay (two of China's most popular online payment services), and receive a password to open the bike's smart lock.

These startups have grown exponentially both in terms of their customer base and the number of cities the services cover. In the autumn of 2017, there were up to 70 bike-sharing brands with 16 million cycles on the streets for a customer base of about 130 million, according to China's Ministry of Transport. However, the stiff competition and thin profit margins forced many to shut down, leaving only leading brands Mobike (orange bikes) and Ofo (yellow bikes) on the streets, which together controlled 90% of the market, according to 2018 news reports just before the merger of Mobike and Meituan.

Meanwhile, China's bike-sharing model quickly expanded into global markets as well. By making urban cycling more accessible, popular, and smart, the new bike-sharing model promised to deliver scaled, sustainable mobility solutions for people and cities around the world. As a result, China's domestic media sang high praise for Mobike and Ofo "pioneering bike sharing in China and the rest of the world". The bike-sharing model was viewed as "a rare and genuine original Chinese tech brand with a sense of design" that functioned like "a new business card for China's innovation in the world".

According to the Sharing Economy Report by China's State Information Center, by the end of 2017, bike-sharing had raised 20 billion RMB (US$3 billion) in financing, reaching more than 400 million users across 304 cities in over 20 countries (see Figure 6.5). In recognition of its transformative contribution to the advancement of low-carbon public transport, Mobike was named among the 2017 Champions of the Earth by the United Nations Environment Program (UNEP).

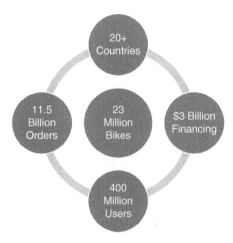

Figure 6.5   China's bike sharing going global in 2017
Data Source: China's State Information Center.

With numerous cars and bicycles "mobilized" (pun intended) on the streets, the phrase "sharing economy" had gone from buzzword to boom market. They were like wheeled flagships of sharing, spreading the gospel of the new economy model. China has entered into the age of "sharing everything" since 2017, and the car-hailing and bike-sharing businesses are only a small part of the broader sharing economy.

Umbrellas, Sex Dolls, and Everything to Share

The year of 2017 saw the most entrepreneurs and capital venturing into the sharing economy, which led to 59 new startups in numerous sectors according to the data from IT Orange (see Figure 6.6). Shared power banks, which had 11 new ventures, ranked first in popularity. These companies offer portable batteries in locations such as shopping centers, restaurants, and other public places. Users rent the batteries by scanning a QR code with a mobile app and pay less than US$.10 (0.5 RMB) for half an hour of use. Shared umbrellas, provided by six new startups, were the second most popular offering.

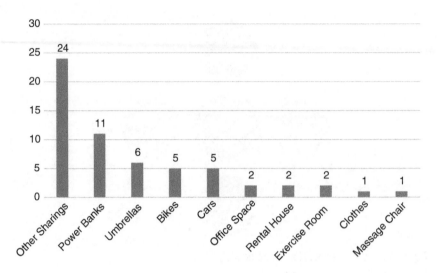

Figure 6.6   Sector distribution of 59 sharing economy startups
Data Source: itjuzi.com (IT Orange, 2017).

Clearly, China's sharing economy has veered away from what the term originally meant: as a peer-to-peer exchange of underutilized goods and services. Unlike Airbnb and Uber, which provide a platform connecting users to underutilized, existing resources like spare rooms and private cars, the sharing economy in China has evolved into something like an internet-enabled rental business, where the sharing companies own all kinds of products that they rent out to users (like bicycles, e.g.). As such, the "other-sharing" category in Figure 6.6 covers quite a few items not easily found in the US and European sharing markets:

- **Girlfriend sex-doll service**. This revolutionary idea stirred up lots of market enthusiasm, but it was shut down quickly in response to public outcry.
- **Capsule beds for sleeping.** This startup's sleeping pods were straight out of a sci-fi film and came complete with a USB port, reading light, and electric fan. But its outlets

were reportedly shuttered due to the lack of relevant safety licenses or permits.

- **Children's toys.** Quality toys for children can be expensive, and kids quickly outgrow them. Why not use them for a few months and exchange them for new toys?
- **Basketballs.** When you walk past a basketball court, you can still stop and shoot hoops even if you don't have your own ball. To rent a ball, users scan a QR code found on basketball-sharing lockers often located near basketball courts.
- **Suitcases**. If you have a tiny apartment, why do you keep large luggage that you only use occasionally?
- **Refrigerators . . .**
- **Luxury handbags . . .**
- . . .

In short, "sharing" in China now refers to almost any short-term rental of a product or service activated by widespread smartphone use and established mobile payment systems. Perhaps "mobile renting" is the more pertinent term (or maybe renting 2.0). But why is mobile renting going so strong in China?

The same key component behind all sharing ventures—from cars and bikes, basketballs, and refrigerators to clothes and massage chairs, luxury handbags, and phone chargers—is the mobile payment mechanism, in which China is by far the global leader. Online sharing (renting) services involve frequent, small payments from changing locations, and the users can simply scan a QR code to complete them with simple taps (the payment systems integrate their smartphones seamlessly with their bank accounts).

Another positive factor is that the Chinese government is open-minded about new sharing-economy businesses. Premier Li Keqiang is a keen supporter, describing the sharing economy as "a reinvigorating force" in China. The economic planning ministry National Development and Reform Commission (NDRC) has issued guidelines to remove barriers to markets for new sharing businesses. Furthermore, municipal governments have also implemented balanced regulations when it comes to controversial issues (e.g. drivers' registration requirements for car hailing). With both central and local governments moving in supportive directions, the sharing economy in China has gained its momentum.

While the sharing economy has moved across the spectrum of personal goods, not all the new ideas will be successful. Most sharing services need to put in substantial upfront capital investments to acquire the shared goods, whether large ones like cars and bicycles, or small ones like umbrellas and portable batteries, but the fee per use must be very cheap to attract the largest possible pool of users. As a result, young companies always need to raise new funding, whereas their earnings and profitability are challenging. For some startups, there is a fundamental question about whether users have a real need to share these products, or whether the goods offered are even suitable for sharing. (**See the "Worst Idea Ever Heard" box.**)

Worst Idea Ever Heard

The umbrella sharing startup E Umbrella was much laughed at for its questionable business model. For a start, compared to a car or even a bike, an umbrella is a lot cheaper, which raises the question whether it is something people prefer to own or even like to share.

E Umbrella launched with an investment of 10 million RMB/yuan (approximately US$1.47 million), charging users 20 RMB (nearly US$3 dollars) for the first time use, plus 19 yuan per umbrella as a deposit, and an additional yuan (US$ 0.07) for one hour's use. Immediately China's netizens questioned the business logic. "The company demands a 20-yuan deposit fee", some commented, "I have a better idea than that—Go buy a new umbrella with that 20 RMB".

Furthermore, E Umbrella seemed to overlook the fact that unlike bikes, umbrellas have to be kept at fixed locations. Bikes can be "docking free" and parked anywhere, but an umbrella needs to be hung on railings or a fence. In practice, E Umbrella's stands were set across the cities near train and bus stations, which limited its convenience value for the broad population.

For the same reason, it was also challenging for the company to ensure that users would return the umbrellas to the stands when done with them. Most users just end up keeping their rentals, or did what they do with their own umbrellas, forgetting where they left the umbrella after the last use. According to news reports, just under three months after launch, E Umbrella had lost almost all of its 300 000 umbrellas across 11 Chinese cities.

Actually, the shared umbrella idea was not from China alone; the same venture idea also appeared in the US market. One startup named BrellaBox pitched a similar concept on Shark Tank, and the venture pitch was called "the worst idea ever heard" on the show by one of the panelists. To be fair, one could easily fill many episodes of Shark Tank with sharing economy ideas competing for the "worst idea ever heard" title.

For example, folding stools were put on the streets of Beijing near bus stops and train stations to be shared. Similar to bikes and umbrellas, the stools also have a QR code on their seats to be scanned before being used. However, they are fundamentally different from bikes and umbrellas in that the goods to be shared are not even "mobile": people don't have to move the stools when using them. Hence, many people reacted with ridicule on social networks: "f I just sat on the stool without scanning, would the company know"?

A unique challenging factor of the sharing economy is that people are still in the process of developing social norms related to shared products and services. Because users are not owners, there is a lack of concern for public resources, and shared products are often abandoned after use. As one commentator observed, "It's human nature not to care". In this, shared bicycles are the most prominent case in point.

The park-anywhere aspect that makes shared bikes so convenient also makes them inconvenient, as careless users often park the bicycles haphazardly along streets and roads, bringing chaos to sidewalks, bus stops, and residential compound entrances. In 2017, more than 16 million of them were put on the streets by over 70 companies. They piled up in the hundreds at heavily trafficked spots, and city governments had to confiscate them in the tens of thousands to clear the way, disposing of them in remote places. City officials also grappled with creative vandalism of the bicycles, ranging from hanging them in trees to setting the entire vehicle on fire.

If the year of 2017 was a year of a "great leap forward", 2018 was a year of consolidation and reflection for the sharing economy in China. Mobike successfully expanded into 170 cities worldwide, owning 7 million bikes and claiming more than 100 million customers; but, at the same time, its losses were also mounting due to:

- **Large capital expenditure**: Typically, sharing economy platforms don't necessarily own the operating assets: Airbnb doesn't own hotels nor does Uber own cars. But the bike-sharing companies had to expend capital to buy and own all the bikes.
- **High operating costs**: The companies need to maintain bicycles and replace the damaged ones, as mischievous vandals smash the locking device, bury them in construction sites, and throw them into lakes and rivers. They need to repair, repaint, and reinstall locks on old bikes, and turn them back into inventory ready for use.
- **Low revenue**: All the bike-sharing companies were providing essentially the same commodity service, which led to an oversupply. Price per ride was very low to start with, and to compete for more market share and active users, companies had to provide various discounts and incentives for riders.

Before falling off a cliff, Mobike managed to change its course by selling itself to a larger internet platform. In April 2018, Meituan Dianping, the largest food delivery platform that had also expanded into car-hailing business, acquired Mobike for nearly US$3 billion to dip its toes into the bike-sharing field. Starting in group discounts and restaurant reviews (hence, it is often described as China's version of Groupon and Yelp combined), Meituan-Dianping sought to build a "super app" offering as many consumers services as possible, from online retail and food delivery to hotel bookings and taxi rides—and bike rides were added in.

Mobike survived (sort of), but its main competitor, Ofo, was not as lucky. In 2018, Ofo was pushed to the verge of bankruptcy as debts to suppliers had come due and user demand for deposits had mounted. A social post in early 2019 by one user said there were over 14 million Ofo users lining up to get their deposits back (normally amounts to 119 RMB/ US$30) before him. "This might be the longest line I have ever been waiting in. . .", the user lamented, and in the comment section, many users suspected that they would never get their deposit money back.

The rapid rise and fall of Ofo shocked the tech investment circle in China because Ofo had successfully completed about 10 rounds fundraising for US$1.5 billion in total in two years' time, with high-profile investors like Alibaba, Ant Financial, Didi, and CITIC in the list. By 2019, many of the bike-sharing startups had declared bankruptcy. Their former fleets of various colors—including the Ofo bikes in bright yellow—were dumped at "bike graveyards" scattered across major cities and their outskirts, forming a decaying monument to an investment craze.

The canary yellow bikes of Ofo quickly disappeared from the streets of Beijing, Singapore, and Paris. But the bikes and the sharing economy are here to stay. Bike sharing may not be a sustainable business model stand-alone, but it is a great service and it may create difficult value when integrated into a platform business, as in the merger of Mobike into Meituan. The bikes have come back to the streets (unexpectedly) during the 2020 coronavirus pandemic, but it is now only a small part of the all-out war in the shared economy—transportation sharing, food delivery, and many more adjacent businesses—by the digital giants.

Meituan, Didi, and Ele.me: All-in-one On-demand Services Platform

After Ofo went bust, bike sharing disappeared from media headlines and venture capital circles. However, the resplendent rainbow fleet of bikes has (again) arrived at the city streets across China—in different colors, of course, representing **Didi, Alibaba-backed Hellobike, and Meituan**. Instead of being run as an independent venture, bike sharing is integrated into the internet giants' empires, complementing their data-dependent ecosystems, as Didi, Alibaba, and Meituan compete to create an all-encompassing lifestyle service platform for everyone. Their war is fighting in many more fields than the bikes, compared to which the brief startup rivalry between and Ofo and Mobike, which cost venture capitalists (only) a few billion dollars, was merely a warmup.

First, a brief history on **Meituan**, the third Chinese internet firm that, after Alibaba and Tencent, reached US$100 billion valuation in the public market. Meituan-Dianping is China's largest O2O platform for lifestyle services, ranging from food delivery, hotel room booking, movie ticket booking, restaurant reviews, group discounts, among other things. It was created through a 2015 merger of two archrivals, Meituan (group discount deals) and Dazhong Dianping (consumer reviews' site); hence, it is often described as China's version of Groupon and Yelp combined.

Within the broad O2O markets in China, restaurants and food catering businesses started early, and their market mode is more mature than other fields. The reasons are simple. First is the sheer number of population: feeding nearly 1.4 billion people is a massive market by itself. Also, eating is probably the most important and frequent social activity for Chinese people, who have a proverb that "people regard food as their prime want" (or, in the Chinese character itself, "heaven"). On top of that, the number of young professionals and workers in the cities is steadily increasing along with the urbanization trend, and they tend to eat out frequently due to their busy lifestyle. For Meituan and Dianping, their core businesses started from restaurant booking and restaurant reviews, respectively.

After a costly price war similar to the ride-hailing startups, Meituan and Dianping merged in late 2015. As a result of the

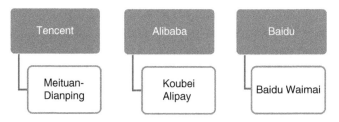

Figure 6.7   The competitive landscape after the 2015 Meituan–Dianping merger

merger, Alibaba decided to sell its stake ownership in Meituan, leaving Tencent as the major shareholder of the combined firm. To stay in the O2O dining service market, Alibaba refocused its own online food coupon site Koubei and the Alipay payment system, which was commonly used in restaurants across the country. In response, Baidu set up Baidu Waimai, a food-ordering and delivery platform based on Baidu's search engine and mapping service (see Figure 6.7).

After the merger, Meituan–Dianping has continued focusing on food-related services, and its close relationship with Tencent supplied invaluable user traffic through WeChat. In fact, a large part of the discussions among WeChat users is about where to book tables at restaurants for group get-togethers and where to find deals. From the core food delivery business, Meituan has constantly expanded into more O2O services, from hotel room booking to movie ticketing, salon, education, sports, and more. (In fact, Meituan founder Wang Xing believes the company's business "has no borders".)

Today, about two-thirds of Meituan Revenue is from the food delivery unit (low margin business), while travel booking (the most profitable) and hospitality services like wedding planning account for most of the remainder. The bike-sharing business from the Mobike acquisition is listed under "new businesses", which suggests a flavor of venture investments. The more business lines Meituan enters, the more directly it competes against Alibaba, whose Alipay has already been widely used by many merchants and shops. In 2018, Meituan-Dianping IPO-ed on the Hong Kong Stock Exchange, and since late 2019, Meituan has reported strong growth and secured its position as China's third largest publicly traded technology company.

Didi, as mentioned, has had stakes in two bike-sharing firms, Ofo (yellow bikes, now defunct) and Bluegogo (blue bikes). The latter was viewed as the third-largest in China, but it was little known because it declared bankruptcy after a little more than one year. In 2018, DiDi bought its stake and relaunched the company by installing a bike-sharing feature within its app to provide services from Ofo and Bluegogo bikes. Although the blue bikes are still cruising the streets, Didi's focus now is its fully controlled Qingju Bike (meaning "green oranges"), marked by the colors of white with aqua green.

Hellobike, whose fleet of bicycles sport a blue-and-white livery, was a latecomer to the bike game, but it quietly pedaled past the sector's early entrants with their operation efficiency. Founded in late 2016, a time when Ofo and Mobike were already leading a cash splurging competition among the bike brands of numerous colors, Hellobike started operating in tier 2, 3, and 4 cities in China—where Ofo and Mobike were largely absent early on—rather than large urban centers like Beijing and Shanghai.

Hellobike is greatly welcomed by users in the lower tier cities, where the city is small, the distance is short, and the public transportation is not well developed. For the residents in smaller cities, bike sharing is often the best alternative to "substitute" bus transportation, which translates more vigorous demand and higher consumer stickiness to Hellobike.

After a wave of bankruptcies of its bike-sharing rivals, the market has opened the opportunity for Hellobike to expand in the larger, tier 1 cities. Meanwhile, it has the strong capital support from Alibaba's affiliate Ant Financial, the world's most valuable fintech company, which is its largest shareholder with an approximately 35% stake as well as the full integration with Alibaba's Alipay app. It is now operating in Beijing and Shanghai, Guangzhou, and Shenzhen, the top four cosmopolitan cities in China.

Figure 6.8   The new competitive landscape of shared bikes

Mobike is now rebranded "Meituan Bikes", and as if dancing on the grave of Ofo, painted in "Meituan yellow" (the color of Meituan's logo) that looks quite close to Ofo's old trademark canary yellow. The new bike war (see Figure 6.8) is not only a battle of capital and scale, but also a battle of efficiency because all the stakeholders have a realistic understanding of the underlying opportunity and challenges. In an unexpected twist, the COVID-19 epidemic led to an uptick in user activity in the first half of 2020. Because of the pandemic, many people prioritized sharing bikes over public transportation to avoid any human contact. To maintain social distancing, they chose to ride bikes directly to their destination without transfer to other means of transportation. However, it remains to be seen whether the short trips covered by shared bikes will eventually, after the world is normalized, be replaced again by buses and subways.

To Meituan, the large database of users of the bike-sharing business is more valuable than its direct financial contribution. (In fact, according to Chinese securities firms' research report, Meituan Bikes is likely to continue loss-making through to 2021 and be a drag on the overall company.) Mobike has been folded into Meituan's all-in-one service app, and it fits well into Meituan's vision to build a one-stop "local services" platform to provide more integrated location-based service (LBS) solutions for its customers. ("**Local services**" is a broad term to include all offline lifestyle services, as in the case of Yelp and Groupon, and on-demand delivery, as in the case of UberEats and Seamless in the Western markets.)

Furthermore, the addition of Mobike is more than a transportation tool; instead, it adds young, healthy, and environment-friendly flavors to its platform brand. The Meituan users need transportation services to get to and from restaurants and other local lifestyle points of interest. With the addition of bikes, they have a new option to travel to bricks-and-mortar restaurants and merchants. The Meituan app aims to be a one-stop shop for everyone's local needs: ordering food or groceries, booking a spa after a movie at the cinema, and taking a Meituan bike to get around.

But the likes of Didi are elbowing their way into Meituan's turf, and their competition will be an all-out war on shared cars, bikes, and much more. Like Meituan, Didi launched a food delivery service across Chinese cities. More recently during the coronavirus pandemic, Didi started daily shopping services in major cities. Families reluctant to shop outside for themselves were able to order groceries and coffee via the Didi app. The drivers would then buy the items and deliver them. Didi has the largest troop of drivers, which enables it to ride into many different services going forward.

The main challenger to Meituan, however, is **Ele.me**, the food delivery arm of the O2O unit within the Alibaba called **Alibaba Local Services Company** (see Figure 6.9). As

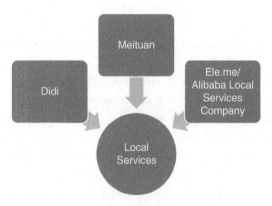

Figure 6.9   All-in-one on-demand services platform—three contenders

mentioned, Alibaba had its own online food coupon and restaurant review site Koubei. In April 2018, Alibaba acquired Ele.me, another market leader in the O2O food delivery market (Baidu effectively quit the scene after selling its food delivery platform Baidu Waimai to rival Ele.me in 2017). To better compete with Meituan-Dianping, Alibaba combined Ele.me and Koubei to form the Alibaba Local Services Company, which received US$3 billion funding from an investor group led by the Softbank.

Ele.me and Koubei have expanded to infiltrate the largely untapped tier 3 and 4 cities, which may have synergies with Hellobike, their sister company in the Alibaba empire. Hellobike is leading bike-sharing penetration in smaller cities and towns, a tactic the company compares to Communist party leader Mao Zedong's "surrounding the cities from the countryside" strategy that led to its civil war victory in 1949. For the food delivery business, the same may apply as China's urbanization push is turning more small-town youths into food catering customers.

The three-way war may last a long time, as both Meituan (plus its main shareholder Tencent) and Alibaba have huge capital power to continue the cash-intensive fight, just like the early price war between Meituan and Dianping (backed by Alibaba and Tencent, respectively). (As a private company, Didi may find it challenging to fight with these deep-pocketed competitors; but in July 2020 the company also announced its plan to go public on the Hong Kong stock exchange.) The difference is that the new players will not just focus on delivery, but also work to digitally power up conventional restaurants (as well as pharmacies, salons, other service shops once firmly rooted in the offline world).

For example, many restaurants found their previous text-only paper menus didn't do justice to their signature dishes. Now, consumers can use apps to scan and browse the menu. Each dish has its own image, so customers can see what they are and their portion size. They can order without needing staff to explain in person. For some restaurants that Koubei has helped to digitize, customers can even play augmented reality-based games to earn discounts on their meals. As such, the O2O players are bringing more "digital persona" to restaurants in the same way Alibaba and JD.com have transformed brick-and-mortar retail stores.

From MaaS to OaaS—a Smart Pie

China's bike-share sector provides the best perspective on the future of China's shared economy. It enjoyed rapid expansion in 2017 and 2018, when startups crammed city streets with colorful bikes, before the bubble burst in 2019, resulting in a wave of bankruptcies and bike graveyards. The rollercoaster ride on two wheels, however, is eventually connected to a new path.

Today, the two-wheelers from Didi, Hellobike, and Meituan arrive at Chinese cities with a new look. Not only do they have new colors, they are also no longer free. (Actually, the price per ride has been trending higher of late.) Fewer bikes are put on the streets, as the companies emphasize more on operation efficiency than pure numbers; hence, the sidewalks are no longer blanketed with discarded bicycles. Meanwhile, they are rigorously adopting AI and Big Data technologies to help them identify the spots with the highest demand for their bikes. Thus, the unused bikes are always where you need them, and they are easier to find than before.

As such, China has not reached "peak sharing" as some suggested. In fact, the consumer demands remain strong, whereas the shared economy has grown up with more mature business models and more advanced technologies. The term "sharing economy" in the Chinese market has numerous synonyms, such as access economy, collaborative economy, freelance economy, gig economy, on-demand economy, and so on. But the startups in this field share the same spirit: using digital technology to eliminate the convenience gap between sharing versus owning, so that people can simply access the things they once had to buy.

Therefore, by removing actual ownership while guaranteeing access, mobile renting is a bigger story—at least for China—than putting existing, under-utilized assets to more frequent use. That may be the reason behind the ultra-broad definition of "sharing economy" used by China's State Information Center for its statistics. According to its definition, China is home to more sharing economy startup unicorns than any other country in the world, and for nearly every two unicorns in China (a total of 83 at the end of 2018 per its calculation), there is one from the sharing economy—an astounding 41% rate of representation (see Figure 6.10).

In China, sharing is a faster and more efficient way to give people a better quality of life, especially when considering that the average incomes in China are still very low and the consumers in many ways are still very price conscious. Furthermore, as the coronavirus outbreak in 2020 started an economic downturn in China for the first time in three decades, the young generation are adjusting their consumption habits with more saving and more sharing.

For example, the young generation's desire to display their lifestyle on social media and the pressure of keeping up with fast-changing trends fuels the need for a fast turnover of their outfits. These conspicuous youth enjoy the status conferred by luxury brands and they do not want to be seen in the same outfit twice on social media, but they need to find access to different outfits within their budget. In 2019, the market saw

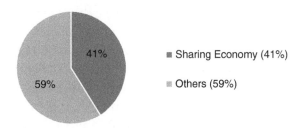

Figure 6.10   Almost half of Chinese unicorns are in the sharing economy
Data Source: China's State Information Center, December 2018.

China's sharing economy extending its reach into fashion and luxury in a big way, and that trend would continue as China's economy begins to slow down.

When people return to work after the epidemic gradually stabilized, Hellobike and other companies found that the long-distance order for shared bicycles—for more than 3 kilometers—has improved significantly. Even for destinations more than "last mile" away, commuters chose to use bike sharing because it allows them to ride in open space without any human contact. As such, a new race has started for shared electric two-wheelers, because bicycles are ideal for trips of between 1 to 3 kilometers, while e-vehicles with motors enable users to travel further distances.

Hellobike is an experienced operator of shared electric bikes, which had previously gone largely unnoticed by its competitors, partly because of its focus in lower tier cities. According to data collected by Hellobike in 2019, nearly 300 million rides are completed on analog bikes every day in China, whereas pedal-assist electric bikes and pedal-free scooters together more than double that number, generating 700 million rides per day. That is more evident in lower tiered cities where the public transportation is less developed. For example, in Nanning, Guangxi Province, the percentage of two-wheel electric vehicle travel exceeds 34% of all transportation, surpassing public transportation and private cars.

Because the e-bike and e-scooter are more expensive than the bikes, their sharing makes even more economic sense. People might be able to afford an e-scooter that costs several thousand RMB (US$1 is approximately convertible into 7 RMB), but if they may leave the city after a year—to start a venture at their hometown, move to another city for job opportunities, or go abroad for advanced studies—why would they buy it? Hellobike and similar platforms offer a new rental model through which people can pay about 200 RMB/yuan (less than US$30) a month to use the scooter.

The electric vehicles involve more technological challenges than traditional bikes. Chief among them is to build safe and convenient charging spots in the cities, which essentially is a battery issue. The national standard for electric vehicles came out in 2019, which set requirements for centralized charging and replacement. Hellobike has put up charging stations since it started offering shared e-bikes in 2017. At these kiosks, e-bike users swap out their flat or faulty batteries for a new one without having to plug in and wait.

Partnering with Chinese battery maker CATL, Hellobike has deployed more than one million electric scooters across China. Meituan, at its quarterly earning call in May 2020, revealed that it is putting "a few" hundreds of thousands of bikes on the streets in Q2, and the initiative will expand further if performance warrants. Around the same time, Didi' Qingju Bicycle reportedly received more than US\$1 billion in financing from both Chinese and foreign investors, and it is also expanding its e-bike-sharing business as its 2020 priority.

Their three-way competition is likely to help e-vehicles soon become a new popular mode of transportation among urbanites. The combination of bikes, e-bikes, e-scooters, and carpooling means that the shared vehicles are gradually overtaking the public transportation and private cars to become mainstream. Travelers see transportation as a service: they no longer need to purchase vehicles; instead, they rent the vehicles by paying for travel services provided by different operators based on travel needs.

As such, a new transportation concept MaaS ("Mobility-as-a-Service"; see Figure 6.11) has emerged in the Chinese cities. MaaS is based on integrating travel services of various transportation modes, powered by IOT (internet of things), Big Data, and intelligent systems that can operate large-scale mobile, smart hardware (vehicles) for optimal scheduling. MasS represents a transformation of mobility: from personal possession of travel tools to the use of travel as a kind of services to consume

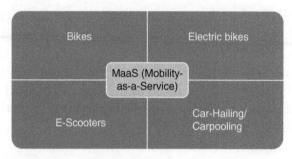

Figure 6.11   MaaS in Chinese "smart" cities

(of course, in a more convenient, green, and smooth way than the traditional bus service).

The MaaS is the backbone of the O2O life service platforms like Meituan, Alibaba, and Didi. Although the mobility services are not profitable by themselves, they engage almost the entire population and collect valuable user data from them; at the same time, they develop critical logistics capabilities in the cities. As mentioned, they are leveraging their armies of delivery workers ("riders" in Chinese) to expand into other businesses like grocery delivery under their own brand names. They continue investing in technologies such as AI powered dispatch system, drones, robotics, and autonomous vehicles to increase delivery efficiency; as such, the virtuous cycle of more powerful delivery system, more business lines, and more users will continue, until they are capable of, as Meituan–Dianping suggests, "delivering everything for everyone everywhere".

As such, the major O2O platforms are the rising "OaaS" (Operation as a Service, for which MaaS is an integral part) providers for offline service merchants. Much like Alibaba that provides smaller businesses digital tools to sell products and compete with larger merchants and brands on its online marketplaces, Meituan and its peers are doing the same for service SMEs (small- and medium-sized enterprises). The O2O platforms are digital enablers, not just connectors of service providers and consumers.

They create affordable infrastructure for startups by allowing them to share the on-demand transportation system and a massive fleet of delivery guys, as well as additional operational systems like smart payment and targeted consumer marketing (in fact, the O2O platforms probably manage the most sophisticated location-based technologies in the market), instead of owning one. When everyone was ordering take-outs during the coronavirus outbreak, Meituan and Ele.me cultivated a clan of "virtual restaurants" that operate only out of a kitchen.

Of course, the "sharing" of the offline operation system is not free, and there is tension rising between the merchants and platforms. The heavyweights charge a significant commission rate, somewhere between 10% and 20% according to media reports; but so far, it has been a win-win for all. In short, the MaaS and OaaS are simultaneously "creating a bigger pie" and "dividing a pie in new ways" among customers, offline business, and internet firms, allowing more Chinese people to benefit from the digital economic boom. As such, the new China Dream is a promising "Smart Dream"—powered by the digital economy innovation—in which everyone gets a piece of a SMARTER pie.

CHAPTER 7

From C2C to 2CC: "Innovated in China"

- Say It with Starbucks
- Premier's Cappuccino Kick-started Entrepreneurship
- Hot on the Heels of Silicon Valley
- 2CC—Panda Kitchen Copied to the United States
- From "Innovative Models" to "Innovative Technologies"
- Peak Unicorn? The Critical Link with the United States

Say It with Starbucks

What will be Starbucks's "next-generation" offering? Whatever that is, it is likely to first come from the China market. In April 2020, Starbucks unveiled a strategic partnership with Sequoia Capital China, a major VC firm, to make co-investments and forge partnerships with "next-generation", "tech-driven" food and retail startups. The alliance will enable the largest global coffee shop chain to get "early access" to innovations and to embed digital technologies across all dimensions of its retail business to harness the power of data-driven analytics, modeling, and decision making, the statement said.

As a matter of fact, only a few years ago, Starbucks was much criticized in China for its late adoption of commercial mobile payments, absence of outdoor delivery, and painful app user experience. If the OMO cases in Chapter 3 demonstrates Chinese new retailers' quest to optimize customers retail experience by integrating online and offline markets with technologies, the evolution of Starbucks in China illustrates foreign companies' initial slow response to Chinese markets' digital revolution and aggressive catch-up of late (see Figure 7.1), as Starbucks adhered to the company's "traditional model" and "core competence" for a bit too long.

Starbucks apparently underestimated the popularity of mobile payments in China. When the company made the first move to permit app payment for the Starbucks gift card in 2015, WeChat pay and Alipay had already spread across the nation from shopping malls in large cities to fruit stands in small towns. Although Starbucks was a pioneer in digital payment in the United States—it launched the first app in North America that enables customers to recharge their membership card with a credit card/pay pal in 2009 and "Order and Pay" app service in 2015—in China, it only introduced third-party mobile payment (WeChat pay of Tencent, Alipay of Alibaba) in 2017.

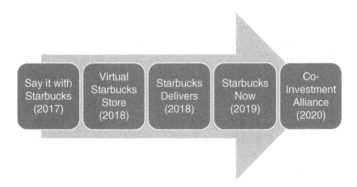

Figure 7.1   The digital path of Starbucks China

Late is better than never. Just before Valentine's Day in 2017, Starbucks shops in China adopted WeChat Pay and partnered with Tencent to launch a social gifting feature *Say It with Starbucks* on WeChat, which was an immediate success. The users can gift a Starbucks drink or digital gift card and add a personal message to accompany the gift using text, images, or video before sending it out. Recipients can save the gifts and memories on their WeChat accounts and redeem their gift at Starbucks stores.

In the accompanied statement, Starbucks claimed that "Say It with Starbucks" was the "first retail brand to bring to life a locally relevant social gifting experience in China". Perhaps, but the reality is, thanks to rapid development of mobile applications, digital gifting had grown in popularity much earlier. This partnership should be a no brainer if one simply considers that WeChat is huge in term of gifting cash, as illustrated by the tens of billions of red envelopes gifted via WeChat during the Chinese holiday seasons. It is not too difficult for Chinese people to embrace the new model of WeChat-enabled coffee gifting—or other kinds of virtual gifting for that matter.

Further, in 2018, Starbucks announced a collaboration with Alibaba's key business lines, including Tmall, Taobao, Ele.me, Freshippo, and Alipay, to significantly elevate the Starbucks experience for Chinese customers. By transcending the traditional limitations of a single app, the *virtual Starbucks store* alliance provides customers a unified one-stop Starbucks digital experience across the Starbucks app and customer-facing mobile apps within the Alibaba ecosystem, including Taobao, Tmall and Alipay. With "Say It with Starbucks", in addition to purchasing a digital beverage gift card, a customer can also add on a physical Starbucks gift, such as exclusive Starbucks merchandise from the Starbucks Tmall flagship store. (See the related OMO discussions in Chapter 3.)

To complement the virtual store, a *Starbucks Delivers* service is added, thanks to the O2O delivery service platform

Ele.me in the Alibaba empire. As discussed in the O2O and Ele.me case studies in Chapter 6, delivery service is a "lifestyle" for Chinese consumers, so the absence of a third-party delivery offering had a detrimental impact on Starbucks sales. The new service captures this opportunity, and extends Starbucks quality-assured, premium retail experience beyond its physical stores.

Once a customer chooses his or her favorite brew, an Ele.me rider would deliver a Starbucks handcrafted beverage within 30 minutes. Building on the positive customer feedback, the program expanded to 2,000 stores across 30 cities within three months after its launch. According to media reports, that pace of expansion and reach is unprecedented in the history of Starbucks. In October 2018, Starbucks also piloted its first "Star Kitchens" within two Freshippo supermarkets (the "digital stores" of Alibaba; see the related discussion in Chapter 3) in Shanghai and Hangzhou. Star Kitchens at Freshippo provide additional scale and reach for fulfillment and delivery of Starbucks beverages and food within 30 minutes, while preserving the third-place experience for customers at Starbucks store locations—under Alibaba's digital roof.

In 2019, Starbucks faced a direct challenge from a startup competitor, Luckin Coffee, which expanded at a fierce pace with a digital tech DNA. Like the Freshippo stores, the Luckin stores are also designed to be "digitalized stores". Most of them are used as pick-up stores: they have limited seating and are typically located in areas with high demand for coffee, such as office buildings, commercial areas, and university campuses— one store in every 500-meter distance (a 5-minute walk) in core areas of Beijing and Shanghai. This enables the company to stay close to its target customers and expand rapidly with low rental and decoration costs.

Consequently, Starbucks in 2019 released a store pre-visit order service called *Starbucks Now* to suit Chinese's on-the-go lifestyle. The mobile order and payment service features GPS functionality to allow users to select their nearest store as well as to customize a beverage, select a food offering, view order time, and pay before their visits to a Starbucks store. (Luckin has been mired in an accounting fraud scandal since early 2020, which will be discussed later in this chapter).

In summary, Starbucks' belated third-party mobile payment adoption was but one example of overseas companies' failing to adapt business operation in China, due to a lack of in-depth understanding of Chinese consumers or an underestimation of the tech savviness of the market. But when the company appreciates the entrepreneurial spirit of the Chinese market, it has quickly learned and integrated the latest digital innovations with its existing model. As the US chain faces increasing competition from homegrown, digital-first players in China, it will continue creating new ways to deepen its engagement with its customers and to deliver new services.

For foreign companies, the Chinese market is not only important for its enormous market size and the fast pace of technology adoption, but also valuable because of the unique innovation arising from it. While the first generation of China tech startups was mostly about the internet and copycat versions of Western sites, the new generation Chinese internet and technology companies are creating a remarkable wave of innovation in business models and product features that challenges the long-held perception of "Made in China" copycatting.

Premier's Cappuccino Kick-started Entrepreneurship

With government endorsement in the background, a dynamic ecosystem of entrepreneurs and startups is organically being built up and rapidly expanding (see Figure 7.2). The network of established internet firms and their seasoned entrepreneurs, endless eager talents, abundant angel investors and venture capital, and a sophisticated manufacturing system are

Figure 7.2   China's innovation ecosystem

collectively making China one of the most interesting centers of innovations in the world.

The government initiatives include building up the innovation clusters, where the administration system is more user-friendly for businesses to get started (e.g. faster corporate registration). Local governments also launch venture funds at province and city levels, whose funding and subsidies help early-stage entrepreneurs to feel more comfortable to take a bit more risk. In dozens of cities like Beijing and Shanghai, university graduates and young professionals can be rewarded with tax cuts in their startup endeavors.

For example, Zhongguancun district in Beijing is the leading innovation center of China and often referred to as the Chinese counterpart of Silicon Valley. The Zhongguancun Science Park, together with numerous high-tech development zones that spread across the country, constitutes a national infrastructure for innovation that is strongly promoted by the central government as well as the local provinces and cities. Four decades ago when China began its reform and opening-up, Zhongguancun started out as a street-selling consumer electronics; today, it is China's AI heartland.

Zhongguancun is filled with angel investors, hedge funds and venture capital firms that have helped fuel the breakneck growth in online and technology-based businesses. It is the home of established Chinese internet players like Baidu, JD. com, and Didi Chuxing; the China offices of global companies such as Microsoft, IBM, and Google; and most notably, numerous startups. During the recent AI boom, Zhongguancun has become the largest and most influential AI innovation cluster in China. While hosting about a dozen advanced AI labs operated by the likes of Baidu, Didi, and Microsoft, it is also the home of more than 1000 early-stage companies specializing in AI and related Big Data analytics.

Also, incubator facilities are established to provide entrepreneurs with inexpensive or even free office facilities, assistance on business services, and common spaces that connect the budding startups with potential investors. They also serve as community centers, where entrepreneurial minds can bounce ideas off each other and startups' owners meet angel investors to present projects. In addition, they are also education centers. Various lectures on entrepreneurship and capital markets are held daily, where experienced professionals share their experiences with the "wannabes".

There are more than 20 incubator facilities (also known as service centers) on the main "Innovation Street" at the Zhongguancun Science Park. Among them, Garage Café, established in 2011, was the first service center at the district where entrepreneurs could work on their projects for a whole day for the cost of one cup of coffee. Today, the 3W Café is by far the most famous one among all the coffee shops (**see the "Premier Li's Cappuccino" box**) because Premier Li had a coffee there with startup owners and wannabe entrepreneurs in May 2015, demonstrating the government's support for entrepreneurship and innovation.

Premier Li's Cappuccino

The "Innovation Street" at Beijing's Zhongguancun Science Park is famous nationwide for edging out other places in attracting the most ambitious and forward-looking entrepreneurs. Young startup owners from all over the country visit the place to find brainstorming peers, business partners, angel investors, and industry backers. But when Premier Li walked into the 3W Café in May 2015, the community received an unprecedented "spiritual boost" on government's endorsement for entrepreneurship.

The vanilla cappuccino that Premier Li ordered has since been referred to as the "Premier's coffee", and the 3W Café has turned into a tourism site. Outside the coffee shop, people wait to take photos against the backdrop of a big screen that shows Premier Li Keqiang drinking coffee with young entrepreneurs.

Near the coffee table where the meeting took place, a wall is now covered with banners calling on people to start their own businesses. According to media reports, at the 3W Café, Premier Li stopped to read the phrase printed on one of its walls: "Life is limited, but *zheteng* is not". (*Zheteng* is a contemporary Chinese term that means "to toil and bustle around something with vague prospects".) "Ah, *zheteng*," the Premier commented, "That really sums up innovation."

Zheteng is a common theme of the internet industry as well as the overall startup community in China. The 3W Café certainly understands the term well: the coffee shop itself is a startup. It is the first crowd-funded café in China, and gradually turned into a multifunctional 3W Space operator that provides café and workplace, integrated with services including fundraising, recruitment, training, communication, and so on to the entrepreneurs. It now runs in over 10 top-tier cities across China, with a total covering space of 40 000m^2.

The core of this ecosystem is a network of "graduated entrepreneurs" from established tech firms at home and abroad. Their BAT or TMD resume provides them immediate credibility in the market. Their alumni network not only provides industry talent to partner with or recruit, but also supplies a pool of seed investors. This development resembles the multiplying effect seen in the Silicon Valley ecosystem the last few decades, where the generations of innovators from Intel, Netscape, Google, and Paypal have created waves of startups.

Equally important, some are former Chinese employees at global technology giants such as Microsoft, Hewlett-Packard, and Yahoo (either in the United States or in their China operations). The trend is cross the paygrades of the major companies as they can replicate the success of the companies they work for, but now as owners rather than employees. Either a senior executive who is financially secure enough to take a major career risk, or a newly minted one who feels he has not much to lose, the aspiration to create a new Alibaba, Pinduoduo, or TikTok is everywhere.

Along with the growing number of "graduated entrepreneurs", there seems to be an endless supply of young and eager talent. Chinese universities graduate more than half a million engineering students a year. At the business schools, more MBA candidates than ever are taking entrepreneurship courses to plan on launching startups (years ago, investment banking and management consulting jobs were the top two career tracks). Another important factor in this booming ecosystem is an unprecedentedly large flow of venture capital (VC)—from both foreign and domestic funds—into startup projects.

Finally, moving beyond "made in China" does not mean the sophisticated manufacturing system is of no use; instead, it is an indispensable part of this ecosystem and a distinctive competitive advantage over other innovation hubs worldwide. China's economy modernization has started with the import-exported related manufacturing. After decades of development, it has a seamless web of sourcing, manufacturing, and logistics services that is second to none. The electronic manufacturing sector is by far the most advanced globally.

With easy access to parts and manufacturing knowhow, this smart hardware hub provides young companies an edge over foreign competitors in terms of production timelines. Also, having the supply chain nearby, early-stage inventors can easily tweak their pilot products at factories, giving them more opportunities to fully develop the finished product. In fact, China's massive electronics supply chains, well established from the "Made in China" era and getting increasingly sophisticated, prove invaluable for the new convergence between hardware and software.

Out of this new dynamic ecosystem, more and more new business models and advanced features are arising in China's internet and tech sectors. For the mobile economy, the Chinese market is poised to be a trend-setter, rather than a trend-follower, in next-generation mobile devices and services. Going forward, as the tech world leaps into "intelligence first" from "mobile first", the Chinese market is setting itself as a leading player.

Hot on the Heels of Silicon Valley

No doubt, the global tech industry's competitive landscape itself is massively changed by the emergence of the internet and tech companies in China. While Silicon Valley of the United States still leads, China is catching up quickly in unicorn births. When people think of Chinese unicorns, they mostly think of the handful of companies that have risen to global status in two decades or less—the initial BAT (Baidu, Alibaba, and Tencent) and the new TMD (Toutiao/Bytedance, Meituan-Dianping, and Didi). Since these few are already valued at more than US$10 billion or $100 billion, they are now known as "decacorns" and "hectocorns". (**See the "Unicorn, Decacorn, and Hectocorn" box.**)

What's exciting about China's innovation ecosystem is that many more Chinese unicorns are coming up behind them. They are still relatively unknown because they are not yet global players; they currently operate only within China, which is large enough for them to reach the unicorn status. According to an October 2019 report from the Shanghai-based research firm Hurun, China has more unicorn startups than any other country in the world. With 206 unicorns, China is slightly ahead of the United States, which the firm says has 203 such tech startups.

Unicorn, Decacorn, and Hectocorn

In medieval lore, the unicorn was a strong, fierce creature—a beast with a single large, pointed, spiraling horn projecting from its forehead—that could be captured only by a virgin maiden. In the myth, the unicorn leaps into the virgin's lap and she suckles it. It is tempting to state that something similar is happening with tech unicorns and virginal venture investors.

In the context of tech investments, a **unicorn** is a privately held startup company valued at over US$1 billion. By referring to the mythical animal, the term was used by venture capitalists to represent the statistical rarity of such successful venture investments. With the same logic, **decacorn** is used for those startup companies over US$10 billion, while **hectocorn** is used for such a company valued over US$100 billion.

By 2019, the search-engine giant Google's parent company Alphabet Inc., together with its peers Apple Inc., Amazon.com Inc., and Microsoft Corp., have become the only four companies ever to reach US$1 trillion market valuation during intra-day trading. (In the 2020 summer, Apple's market capitalization reached $2 trillion.) The biggest tech companies have soared to heights once unimaginable, and for now, there is no "X-corn" term developed for the US$1 trillion threshold.

In 2013, Aileen Lee, the founder of Cowboy Ventures, coined the word for 38 exceptional startups. Currently, there are more than 400 unicorns globally (among them many decacorns and hectocorns). The rapid increase in the number of unicorns has led to many questioning whether the valuation of tech startups is sustainable. In the end, the word *unicorn* betokens something both rare and wonderful.

Notably, the Hurun report counts more Chinese unicorns than other research firms. CB Insights, for example, recorded 99 unicorns in China in the same October 2019. The vast majority of the unicorns that are listed by Hurun, but not by CB Insights, is relatively small, with valuations between US$1 billion and $2 billion, which are likely to be those Chinese local startups below the global market watchers' radar. Of course, the valuation of private tech companies is somewhat vague—especially on the borderline of US$1 billion, but the indisputable fact is that China is challenging the United States to become the world's biggest hub for unicorns. (Together,

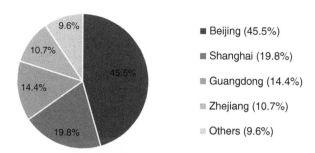

Figure 7.3   Regional distribution of tech unicorns in China chart
Data Source: CNNIC, December 2019.

the two countries host more than 80% of the world's unicorns, according to Hurun.)

For decades, the United States has remained at the core of delivering new innovations in technology, but Chinese unicorn creation has lately been hot on the heels of Silicon Valley. In addition to the Zhongguancun District in Beijing, the cosmopolitan city of Shanghai is also a top city for new unicorns and VC investments (see Figure 7.3). In the southern province of Guangdong, the city of **Shenzhen** is arguably the best place for hardware startups. Once a small fishing village next to Hong Kong, Shenzhen, after three decades of development, is now home to domestic tech giants such as Tencent, Huawei, and ZTE, and it has become an electronics manufacturing center for the global tech industry.

Hangzhou City (in Zhejiang Province) only joined the rank as a tier 1 innovation center city in recent years, but its momentum is strong. It is where Alibaba headquarters resides, and there is a robust startup ecosystem emerging along with the development of the Alibaba empire. Currently, Alibaba Cloud, the cloud-computing unit of Alibaba, is one of the fastest-growing units and poised to be the main business of the group. Corresponding to that, the famous West Lake (Xihu) district created a "Cloud Town" (Yunqi

Xiaozhen) with Alibaba to cultivate startups and research in the cloud industry.

Meanwhile, Alibaba's cloud product known as City Brain is used by the city of Hangzhou. The system gathers traffic information across Hangzhou, such as videos from street intersection cameras, GPS data on the locations of cars and buses as well as environmental conditions. Then it uses artificial intelligence to analyze the data and coordinate more than 1000 key traffic lights and road signals to guide real-time traffic flow. The City Brain did reduce road congestion in Hangzhou, one of China's most traffic-clogged big cities, and it is beginning to test city energy system management.

Furthermore, new hubs are also emerging in southwestern China, far away from the leading coastal cities previously mentioned. **Chengdu**, capital of Southwest China's Sichuan Province, set up a special Big Data department to promote the development of the city's Big Data industry. Just like Hangzhou, Chengdu has actively used Big Data to manage city traffic. Interestingly, based on the Big Data research by CNB Weekly's Rising Lab project, Chengdu is the immediate next city to join the tier 1 city club (after the long-established group of Beijing, Shanghai, Guangzhou, and Shenzhen).

Guizhou, the capital of Guizhou province, is another Big Data city. Guizhou used to be one of the country's poorest and most underdeveloped provinces, and it was renowned for tobacco, tea, and moutai, the country's most famous liquor. From the data industry perspective, however, Guizhou occupies a plateau surrounded by mountain ranges and has a mild climate; it is also rich in water and coal, allowing electricity to be generated cheaply and abundantly. In 2016, the province became China's first "Big Data Comprehensive Pilot Zone", placing it at the forefront of the country's growing Big Data industry. Since then, many large data center projects are being developed.

Anti-9-9-6 Movement

GitHub (owned by Microsoft) is the world's largest open-source site that enables programmers to collaborate on coding. In late March 2019, however, it was used by a group of Chinese software developers for an activism project called "996.ICU", a reference to the "9-9-6" work schedule—9 a.m. to 9 p.m., six days a week—and the likelihood that one eventually ends up in the intensive care unit because of it. The initiative calls on companies in China "to respect the legitimate rights and interests of their employees", and it quickly led to a nationwide debate.

Chinese entrepreneurs routinely work 9-9-6 or 12 hours per day, six days per week. At China's leading innovation hubs of Beijing, Shanghai, and Shenzhen, it wouldn't be a stretch to see startup teams working 10-10-7 or 12 hours every day of the week. In the "really" busy season, that could turn to 10-2-7 (for the avoidance of doubt, here "2" means 2 a.m.). In a way, China's 9-9-6 work culture almost makea Silicon Valley look too relaxed, because the US West Coast innovation center is more accustomed to 9-5-5.

The discussion gained momentum as users added to a blacklist of more than 150 companies that push their staff to work excessive hours, posting evidence of unpaid, often compulsory or heavily encouraged overtime. The debate spread across Chinese social media, where many posts criticized the tech industry's work culture as "inhumane". Even Chinese state media weighed in. The state-run *People's Daily* said in a commentary: "Employees who object to 9-9-6 cannot be labelled as 'slackers'. Their real needs should be considered." The anti-9-9-6 movement even received support from the employees at Microsoft, who have launched a petition calling on the company to stand behind Chinese tech workers.

Nevertheless, Alibaba founder Jack Ma, who is among China's first generation of internet entrepreneurs, claimed himself as a proponent of the 9-9-6 work hours, calling it "a huge blessing" and "an honor" rather than a burden. And those who worked longer hours will get the "rewards of hard work", he said. Mr. Ma's comments prompted criticism from Chinese social media users. "Did you ever think about the elderly at home who need care, [or] the children who need company"? wrote a Weibo user in response to Ma's post. "If all enterprises enforce a 9-9-6 schedule, no one will have children", some added.

The Chinese unicorns are growing fast, in part, because their leaders feel they don't have time to grow slowly. Like their peers everywhere, they emphasize speed and rapid, continuous improvement. And the Chinese market may be the most competitive in the world. (**See the "Anti-9-9-6 Movement" box.**) As many cases in this book illustrate, the young companies release products before they are fully tested, change their business models quickly as they gather customer data and market feedback, and do not hide their ambition to become champions of their market sectors. Their incredible pace almost makes Silicon Valley look sleepy.

2CC—Panda Kitchen Copied to the United States

At the earlier years of internet, the first generation of China tech companies had their start as copycat versions of Western sites, leading to the "Copied to China" (C2C) perception long held by global markets. Now, relating to business models and features, the new trend is "to China Copy" (2CC), that is, Western companies copying new business models that are coming out of China (see Figure 7.4). For example, the free parking bike-sharing business was started by Chinese companies Ofo and Mobike, leading to copycats such as Spin and LimeBike in Silicon Valley.

More recently, shared food preparation services blossomed in China, and in 2019, former Uber CEO Travis Kalanick launched his new startup CloudKitchens, a kitchen-sharing concept for restaurants and take-out orders, which followed the successful expansion of shared-kitchen companies in China such as Panda Selected (Interestingly, Kalanick is partnering with the former COO of the bike-sharing startup Ofo, Yanqi Zhang, for CloudKitchens. Ofo, as covered in the previous chapter, went into bankruptcy after a costly price war with bike-sharing peers.)

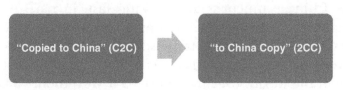

Figure 7.4   From "C2C" to "2CC"

Panda Selected is the leading player of shared kitchen model in China—kitchens that invite restaurants to share the space to focus on take-out orders. Part of the O2O food delivery value chain, the shared kitchen caters to a young on-the-go population that often orders food by mobile app and gets quick take-out deliveries. The shared model lowers the cost of doing business for commercial restaurants—no need for large dining areas or kitchens that serve just one restaurant. Panda Selected was started in 2016 and it already had more than 120 locations in China's major business hubs by 2020.

Part of the reason for the rapid growth of shared bikes and kitchen services is that China has the largest mobile-only population in the world. 900 million smartphone users are connected by the same language, culture, and mobile payment system. The implication is profound: new mobile applications probably can receive market feedback and achieve meaningful scale more quickly in China than elsewhere because new technology based on "human infrastructure" tends to spread faster in China.

Furthermore, Chinese companies seem more inclined than their Western counterparts to migrate across sector boundaries and create larger ecosystems. Because of the "core competence" doctrine that has governed corporate strategy thinking in the West, most foreign multinational corporations tend to focus on what their key business fields are and avoid expanding across diverse sectors. In contrast, the main internet and tech players in China all have an expansive vision to operate networks of businesses that can support each other and supplement each other's capabilities, into "platforms" where consumers can immerse themselves for every possible consumption need.

The rise of China's super platforms is driven by both consumers and mobile payments. For consumers, it is easier to have a single app that offers several services or combines related activities and services. For example, a mobility app could combine hailing taxis, ride sharing, and bike or e-scooter rental.

In e-commerce, previously separate areas such as entertainment, food, and general merchandise are combined into online marketplaces. For the companies, the mobile payment can enable their platforms catering to a variety of consumers' needs such as food ordering, travel bookings, premium video watching as well as (naturally) financial services like small loans and wealth management.

For instance, many have thought of Alibaba as China's version of eBay and social media company SINA Weibo as China's version of Twitter. In fact, these firms have later transformed their business models fundamentally. As seen in the case studies of this book, Alibaba is more like a mix of Amazon (e-commerce), eBay (online marketplace), PayPal (mobile payment), and Netflix (video and filming), whereas Sino Weibo is like a media platform combining Twitter (microblogging), Instagram (picture sharing), and YouTube (video sharing).

Hence, the market sees China's side ahead of the game in entwining the worlds of social media, online entertainment, and e-commerce—even for the "super platforms" in the United States. Compared to their US counterparts, Chinese mobile apps are more advanced for content, social and commerce, and they are more transaction-based. In the super competitive Chinese market, the "cloned" apps have evolved quickly and pioneered new concepts, meanwhile it's often the US "original" companies that are the ones late to the game of new features (see Table 7.1).

Thanks to advanced data analytics, Chinese tech companies are turning into business ecosystems in which sectors that once seemed disconnected are now integrated seamlessly by user data. The super-platforms have an enormous advantage from what economists call "network effects"—the more services they provide, the more people use them; the more indispensable they become, the more user data they keep getting. Their Big Data on users is a rich source when they pursue new market sectors. The result is an ecosystem with both physical

Table 7.1   New ideas from China to the United States

United States	China
Facebook, in 2019, planned to integrate private messaging, groups, and payments with **Instagram** and **WhatsApp**	**Tencent's WeChat** pioneered this years ago
Google, in 2019, added new shopping features to YouTube to give shopping recommendations, share affiliate fees, and enable brands to include shoppable ads	After its 2016 acquisition of video-streaming site Youku Tudou, **Alibaba** has placed advertisement from Taobao and Tmall vendors within the online videos, with links to mobile payment ("buying while viewing")
Instagram, in March 2019, launched a beta shopping feature that enables buyers to click on style and beauty items and purchase directly from the app	Chinese social shopping app **RED** has long offered that as a built-in feature
Amazon's livestreamed shopping feature **Amazon Live** was launched in February 2019, which features livestreamed video of hosts demonstrating products that viewers can buy directly from a carousel that displays under the video	**Alibaba**'s marketplaces offered the same feature years ago
Facebook, in November 2018, launched **Lasso,** its short video app designed to compete with TikTok; in Brazil, Instagram launched TikTok-like video-music remix feature **Reels**	**TikTok,** the short video app from Chinese company ByteDance, integrated KOL (key opinion leaders), short videos, and e-commerce earlier

and online businesses that provide one end-to-end, data-driven consumer experience from consumption idea to service delivery. Increasingly, the platforms predict the needs of customers before they are articulated.

On learning from Chinese innovation, Travis Kalanick and the Uber company are actually in agreement. While Kalanick is replicating Panda Kitchen, Uber has set itself to be the "operating system for your everyday life", which sounds like a super platform in China's digital economy. The company has merged Uber Eats, the company's food delivery service, into the main ride-hailing app. The app also adds bus and train schedules and public transportation directions, and users can buy bus and train tickets through the app, as the company said that it's trying to be the "Amazon of transportation."

The truth is, after the new overhaul, Uber's new look is more like China's Meituan Dianpin (see the case study in Chapter 6). A combination of Groupon, Yelp, Seamless, Uber, and more, Meituan is a one-stop service platform for all local needs in a Chinese city: food delivery, restaurant deals, movie tickets, or hotel bookings—and it also provides bikes, e-scooters, and car hailing services for you to get around. Meituan turned profitable for the first time in late 2019, whereas Uber is still in deep loss. For Uber, adding public transportation information is not expected to bring in revenue; the hope is that if more people are opening the Uber app more often, the company may be able to monetize that one way or the other. (That also sounds familiar if you are used to Chinese internet platforms' thinking.)

Nevertheless, Alibaba, Tencent, Meituan, and Weibo could still be viewed as Silicon Valley models that are "evolved in China"—not examples of Chinese companies' innovative prowess in technology. But that is only partially true. During the last few years, Chinese tech companies have proven their mettle by catching up to global rivals in the smartphone and fourth-generation (4G) technology development process. As the following sections show, going into the era of 5G and iABCD, a new generation of tech startups is emerging, which can no longer be simply described as the Chinese versions of Silicon Valley firms.

From "Innovative Models" to "Innovative Technologies"

The new generation of Chinese startups represent both technological innovation and business model innovation. However, unicorn company leaders these days put more focus on technological innovation to gain an edge. To be sure, new business models continue to be innovated in China—more rapidly than anywhere in the world—to improve user experience. But business model innovation is more commonplace; unicorns increasingly seek to distinguish themselves

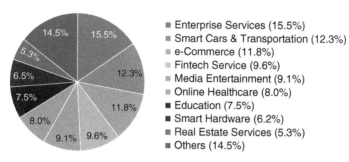

- Enterprise Services (15.5%)
- Smart Cars & Transportation (12.3%)
- e-Commerce (11.8%)
- Fintech Service (9.6%)
- Media Entertainment (9.1%)
- Online Healthcare (8.0%)
- Education (7.5%)
- Smart Hardware (6.2%)
- Real Estate Services (5.3%)
- Others (14.5%)

Figure 7.5   The sector distribution of internet and digital tech unicorns
Data Source: CNNIC, December 2019.

through unique intellectual property, so-called "deep tech" and "hard tech".

Such a shift of focus, from "innovative models" to "innovative technology", from "online platform building" into "digital technology research" is reflected in the distribution of unicorns. According to CNNIC data at the end of 2019, while consumer-oriented e-commerce and media entertainment unicorns are still an important part, about half of the unicorns are in the fields of enterprise services, transportation, fintech, healthcare, education, and smart hardware, which are rapidly implementing 5G-powered iABCD technologies into traditional industries (see Figure 7.5).

Autonomous vehicles, for example, is a sector with multiple interesting players. China is striving to gain leadership in developing next-generation cars with autonomous driving and internet connectivity. The year 2018 saw a foundation for Chinese autonomous driving rules put in place, which will give a big push for tech deployment going forward. For instance, the city of Hangzhou is planning on an autonomous vehicle-compatible "super-highway", which is expected to be completed by 2022. **Baidu** is selected by China's Ministry of Science & Technology as the "national champion" in self-driving cars. The self-driving vehicle unit, called the Apollo project, has secured partnerships with car companies such as Ford, Volvo Cars, and BMW.

In the **smart hardware** area, China boasts a few global leaders, powered by the "Made in China" manufacturing prowess. For instance, **DJI**, the world's leader in civilian drones and aerial imaging technology, has two-thirds of the world's drone market and a global workforce of more than 10,000 employees. But DJI still has competitors in China for specialized areas: in the field of agricultural UAV (unmanned aerial vehicle), Guangzhou-based **Jifei** technology is the leading manufacturer.

Education (so-called edtech) is an especially booming section, which remains a bright spot of investment activity after the COVID-19 breakout, as the pandemic becomes a catalyst for educational institutions worldwide to search for innovative solutions. Language learning, along with early literacy and math, are the areas in which educators generally agree that AI can deliver positive learning results. Chinese families have a long history of investing in their children's education for their future, and after-school tutoring is estimated to account for about 15% of household income in modern China, making it a giant market with tremendous growth potential. Compared to traditional brick-and-mortar schools, online classrooms have a broader student base and lower operating costs.

As such, a number of successful startups in the education sector has added AI to the traditional education model and it has quickly taken off. Among them, **VIPKid**, an online classroom for English learning, is probably the most funded and best established. VIPKid provides students aged 5 to 12 in China with access to English-language teachers in North America, via one-on-one video sessions, at a level that matches top-tier schools in the United States. In 2018, VIPKid developed a partnership with Microsoft to implement Microsoft's AI technology to further improve the learning experience.

Its competitor, **LAIX** Inc, which means "Life empowered by AI to reach (X) infinite possibilities" according to its founder, has developed the English learning app Liulishuo ("speak fluently" in English) with AI built in. The app records users speaking, and then analyzes what they have said and delivers lessons based on areas in which they can improve—pronunciation, grammar, and vocabulary. Leveraging massive volumes of user data, the app continuously evolves and delivers more personally tailored learning programs to each user. Moreover, it integrates games and social sharing into the platform to make for a more fun and interactive learning experience.

Another interesting player in the field is **Zuoyebang** ("homework help" in English), which had its origin in "image search". It was built to be a Q&A platform for middle school and primary school students. Zuoyebang app allows users to search for homework answers and one-on-one help by uploading photos of their problems. The app employs computer vision technology to read the questions from the image and then makes a quick search in its homework database.

But soon all of them, as well as their competitors, will converge on developing online courses where AI software can monitor and improve the users' learning process. The algorithm tracks learners to ensure they remain engaged in their studies. With tutoring sessions recorded on file, the AI algorithm will look specifically at students' facial reactions to the material they learn—especially students' eyes and how they move. Personalizing education with an adaptive algorithm may even be able to replicate that one-on-one environment without the need of a human instructor. For English studies, for example, AI technology has already incorporated the key characteristics and core functions of high-quality, effective teachers:

- **Hearing.** The AI system hears through a speech recognition and scoring engine based on deep learning technology. It can address uncertainties introduced by background noise as well as different speaker accents and proficiency levels.

- **Evaluation.** Using its natural language processing, or NLP, capabilities, it can evaluate users' speech based on several criteria, including pronunciation, vocabulary, grammar, fluency, and coherence.
- **Feedback.** Based on the above evaluation, the AI teacher provides various forms of real-time personalized feedback to users through an intuitive interface.

In April 2020 – the peak of the coronavirus pandemic in China, online education startup **Yuanfudao** raised US$1 billion from its Series G funding round. The transaction pushed its valuation to US$7.8 billion, according to a company statement, making it one of the most valuable ed-tech companies in China. Fueled by the shelter-at-home policies, China's online education businesses have emerged as one of the growth areas in an economy that has been hit hard by the novel coronavirus outbreak. As is true in every section, competition escalates. The newcomers are, again, the short video platforms.

Chinese video apps like Douyin, Kuaishou, and Bilibili are mostly recognized as entertaining virtual places for people to goof off, but now they are taking steps to embrace the education sector because the kids and teenagers are already their users. The young learners spontaneously upload quality cross-topic videos to enrich the platforms' content ecosystem, ranging from finance to computer science, foreign language to software skills. Interestingly, compared to traditional online education platforms, the video platforms create more of a "community culture" for studies.

Bilibili, whose signature feature is "bullet comments" (a user conversation system where messages are directly displayed on top of the video as it plays), is now using the same feature to foster a "studying together" environment. It is not clear whether the parents would buy that, but this is, after all, China, where no combination of new technologies and business models will pass without first being tested.

Peak Unicorn? The Critical Link with the United States

Of course, not every Chinese unicorn will succeed. Saving a few that truly possess proprietary, unique technology, the design and manufacture of unicorns has been industrialized, with the central ingredient being online service, supported by mobile apps and marketed by social media. Globally—not only in China, but also in Silicon Valley and emerging innovation hubs like India—all markets have tailored themselves to churning out such beasts, using the same ingredients and adding in local characteristics. When the term unicorn was coined in 2013, only 38 unicorns were identified; yet, by early 2020 the market data firm CB Insights counted more than 400 of them, more than a tenfold increase from 2013.

The rapid increase in the number of unicorns has led to many questioning whether the valuation of tech startups is sustainable. Operationally, they face numerous challenges, particularly as their losses widen with growth and their need for talent and other resources scale up. A close inspection at the herd in the stable is truly worrisome: some seem bred for show, not for labor; some have gathered too much fat to win their races; even worse, some are sick from their own behavior problems. Then the 2020 coronavirus pandemic came to ravage the herd.

As such, the years 2019–2020, when WeWork failed to work, have become the inflection point where the effervescent mood starts to deflate. **(See the "Hi Ho! Hi Ho! It's Off the Cliff We Go!" box.)** WeWork, whose global meltdown is detailed in the box, also had a Chinese episode. WeWork China, presently 59% owned by WeWork, added outside investors as well. In 2018, only a year after WeWork China launched, Softbank, along with Temasek, Turnbridge Partners, and others invested in the second of two US$500 million rounds, valuing the unit at roughly US$5 billion. Given that WeWork is close to bankruptcy in 2020, the valuation of WeWork China is likely also well below the US$5 billion mark at which the institutional investors invested.

Hi Ho! Hi Ho! It's Off the Cliff We Go!

Looking less like the unicorn tapestries and more like the Brothers Grimm, the story of WeWork and its pulled IPO is a fairy tale without a happy ending for its investors, although the SIFs have largely managed to escape the reputational hit taken by Vision Fund. In the tale of Snow White, the seven dwarves return from their shared workspace one day to find their beloved Snow White asleep under a spell broken only by the kiss of a handsome prince. Unfortunately for the investors, the CEO did not turn out to be the handsome prince, and the company never cast off the evil spell after filing for its IPO.

Fueled by the magic of endless capital, WeWork transformed the workplace in cities around the world, becoming the largest single tenant in New York City in the process. Its principal backer, Vision Fund, injected billions of cash into the phenomenon. Although not truly a "tech" company, at a valuation of US$47 billion, WeWork and its founder, Adam Neumann, benefited from the fairy tale rise of the tech unicorns.

Then everyone woke up when the IPO was filed in August 2019, and the fairy tale was over: it turned out that Adam Neumann was no handsome prince. According to the *Wall Street Journal*, during the IPO process, Neuman and friends chartered a private jet to Israel, smoking marijuana enroute and leaving a cereal box crammed with the substance on board for the return flight. Reports of tequila shots at company meetings and the purchase by the loss-making company of a US$60 million corporate jet also contributed to the questioning of his management style and judgment.

In October 2019, Neumann was ousted as CEO (with a "happily ever after" payout of over US$1.7 billion). Masayoshi Son and the Vision Fund took over the company at a valuation of US$8 billion to avert financial disaster and the IPO was "postponed indefinitely". The carnage did not stop there. Soon the advent of the coronavirus pandemic delivered a deadly blow to the shared workplace model. When Vision Fund took the position that the conditions to the buyout deal had not been satisfied, thereby permitting it to walk away from the commitment without penalty, it became the target of a March 2020 lawsuit by the board of the (likely former) unicorn.

Like WeWork, many Chinese unicorns' valuation has started to shatter in the pandemic-led economic downturn, but Luckin Coffee, the (once) competitor of Starbucks, probably is the most dramatic one. As mentioned at the beginning of this chapter, Luckin applies an internet mindset to its business model. No cash is accepted in its outlets, and customers can only order and pay through the Luckin app without even speaking to anyone during the whole process. In its May 2019 IPO document, the company described itself as a coffee "network" and prided itself on offering a "100% cashier-less environment." That slapped a new-economy aura on a largely routine business, not unlike the office-sharing company WeWork.

Luckin's meteoric rise was the well-trodden path taken by on-demand internet service providers in China—to quickly obtain critical mass customers by providing steep discounts/ subsidies, then leverage the customer scale for capital market fundraising, followed by further expansion to squeeze out competitors, the so-called "burn cash-for-market" in model. The company raised more than US$2 billion in total from private and public investors, but burned through a lot of the cash by expanding rapidly and doling out heavy subsidies such as two-for-one discounts. Skepticism swirled around the sustainability of Luckin's business model well before its Nasdaq listing, but the company continued growing at a blinding pace (and loss). In early 2020, Luckin said it already surpassed Starbucks to become China's largest coffee chain by number of stores.

Then, on April 2, Luckin announced that many of its sales—as much as 2.2 billion RMB in sales transactions (US$310 million)—had been faked and some of its costs and expenses were also substantially inflated. When the US stock market opened, Luckin shares plunged as much as 81% on the immediate trading day. According to the *Wall Street Journal,* Luckin sold vouchers redeemable for tens of millions of cups of coffee to companies that had ties to Luckin's chairman and controlling shareholder. Their purchases, procured partly by a fictitious employee called Lynn Liang, helped the company book sharply higher revenue than its coffee shops produced (see Figure 7.6).

The scale and audacity of deception has stunned international investors and confounded regulators, both in China and the United States. In fact, the Luckin Coffee case also illustrates that China's innovation ecosystem is not purely of China; instead, its ties with the West, especially the United

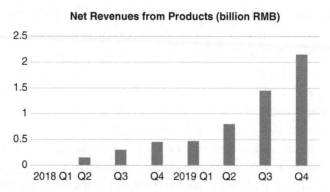

Figure 7.6   Luckin Coffee's dark brew revenue booking
Data Source: SEC filings (1 billion RMB = US$140 million).

States, are critical elements that make the ecosystem dynamic and robust. Little known by people outside of the venture capital community, the United States has played an outsize role in nurturing China's technology sector. Adding to the ongoing tech war, the caffeine shock of Luckin Coffee (and similar cases) have profound impact on China's innovation ecosystem going forward.

First, venture capital from the West. Even before unicorns like Luckin were going public in the United States, they have attracted venture capital from the United States and other Western countries. Luckin's main shareholder is the venture fund Centurium Capital, whose capital mainly comes from institutional investors, including the Washington State Investment Board (United States), the Ontario Teachers' Pension Plan (Canada), and Singapore's sovereign-wealth fund GIC and Temasek. Many of China's most successful tech companies, from Alibaba and ByteDance to Meituan and Xiaomi, have benefited from US venture capital helping to prop up their growth. (In early 2020, the Centurium Capital fund was on track to close a US$2 billion fundraising for a new fund, for which the Washington State Investment Board held back its commitment after the Luckin disclosure.)

According to the early 2020 research report from Rhodium, US funds and investors account for almost one-fifth of the venture capital raised since 2000 by Chinese companies. When the money from passive American investors like college endowments, pension funds, and retirement funds (such as the Washington State Investment Board, which acts as a LP investor in venture funds) is added to the direct investments of venture capitalists, that ratio would get to more than a third. These days the US lawmakers from both parties have discussed new legislations that would make it harder for Americans and their pension funds to invest in Chinese companies.

Second, foreign capital in the IPO market. The United States is a popular choice of jurisdiction for Chinese unicorns to go IPO—even more so than in China itself. By the end of 2019, China had 135 internet and tech companies that were listed in mainland China (Shanghai and Shenzhen, 50), the United States (54), and Hong Kong (31) (see Figure 7.7). It is remarkable that the percentage of US-listed companies (40%) actually exceeded that of domestic-listed Shanghai and Shenzhen. After the Luckin IPO, many retail investors and asset management firms bought into its growth story, including US mutual fund giant Capital Group and Blackrock as well as hedge funds such as Point72 Asset Management and Lone Pine Capital.

Figure 7.7   The United States is the No.1 listing choice of Chinese unicorns
Data Source: CNNIC, December 2019.

Luckin's fall has rekindled long-running tensions over the US Securities and Exchange Commission (SEC)'s inability to inspect financial records of Chinese firms to protect American investors. After all, it's far from the first time that a US-listed Chinese company has been embroiled in allegations of accounting manipulation. The United States and China have a memorandum of understanding signed in 2013, allowing Chinese companies to withhold information if their local laws forbid them from sharing it. The SEC has been locked in a decade-long struggle with the Chinese government to inspect audits of US-listed Chinese companies. In April, after the Luckin Coffee fiasco, SEC chief Jay Clayton warned investors about disclosure problems with Chinese companies.

For some US lawmakers, that accounting waiver should probably be reviewed as to whether it is still appropriate, and if not, be rescinded. Quite a few bills have been proposed. Some would force Chinese companies to abide by federal auditing rules and disclosure requirements when they seek US stock exchange listing. Some aim to mandate the delisting of foreign companies that fail to comply with auditing regulations. The consensus is to force Chinese companies to be more transparent or risk being delisted. (In the summer of 2020, after losing more than 90% of its peak market value, Luckin was ultimately delisted from the Nasdaq exchange.)

Then, in August 2020, the SEC launched an investigation into iQIYI, the video platform often labelled the "Netflix of China" (see the related discussion in Chapter 4), after due-diligence firm Wolfpack Research found instances of allegedly exaggerated sales and revenue in a report published in April 2020. Meanwhile, the China Securities Regulatory Commission (CSRC), the Chinese counterpart of SEC, sent US authorities a new proposal about co-auditing Chinese firms. "Resolving issues of common concern through dialogue is the only way to achieve win-win results", said the CSRC, clearly not pushing for a decoupling.

Third, expansion and acquisition in the United States. The US market is highly significant for Chinese tech companies to go global, both for its vast size and dynamic innovation. If Luckin Coffee had not been derailed by the accounting scandal, its next step after the Nasdaq listing would have been to expand into the US market and compete with Starbucks on the latter's home court. Along with its expansion, it would most likely acquire US companies to accelerate its growth.

In early 2020, new rules developed by the Trump administration to empower Committee on Foreign Investment in the United States (CFIUS), the agency that screens foreign direct investment for national security risks, became effective. The authorities of the President and CFIUS are broadened to review and block China (and other foreign capital) from investing in US companies involved in critical technology, infrastructure, and data, which is discussed in Chapter 9.

Ultimately, impeding the cash flowing between China and the United States, two leading innovation centers of different strengths, will do more than just harm commercial returns, it will sever the flow of ideas and innovation. No doubt, the China–US tension adds extra challenges to the Chinese unicorns, at a time the market breakdown triggered by the coronavirus pandemic is depressing unicorns' growth and severely testing some of their business models. For many younger unicorns, the year 2020 could be their first direct experience of how cycles really operate, and how a local startup business could be disrupted by geopolitical swings.

Of course, many of them will survive and thrive. Compared to the "copycatting" generation, the new generation of Chinese companies is more innovative in products and technologies for local market needs, more willing to accept outside investors, and has a more global outlook from the very beginning. Backed by the emerging ecosystem of entrepreneurship in China, the story of China is rapidly transforming from the old "Made in China" factories to the younger "Innovated in China".

"We are excited to tap the tremendous energy of technology entrepreneurs from two of the world's largest and most dynamic markets, to pioneer innovative solutions that could reimagine the global retail landscape," said Kevin Johnson, the CEO of Starbucks, on the 2020 partnership with Sequoia Capital China. Indeed, his company's deep involvement with the Chinese market may be a harbinger of more foreign companies that will come to China to look for new markets, features, products, and business models.

CHAPTER 8

Land of Big Data and Its Legal Framework

- Facebook and China Awaken to Privacy
- World's Largest Data Worker Pool in the Land of Big Data
- From Human Flesh Search to Big Data Killing
- The AI Age: Privacy, Security, and Value
- New Rules for a New Game
- Ride-hailing Sex Incidents: More Questions than Answers
- Civil Code 2020 and More to Come

Facebook and China Awaken to Privacy

When Mark Zuckerberg testified before Congress in early 2018 on Facebook's data practice, he warned that regulating the platform's use of personal data would cause the United States to fall behind Chinese companies when it comes to data-intensive innovation, such as AI. His argument reflected the conventional wisdom that the Chinese internet industry has a tremendous amount of user data accumulated for AI research thanks to lax regulation on data collection in China. In other words, Chinese companies will have an edge if US companies like Facebook are constrained by data protection regulation.

Zuckerberg's argument seemed to be echoed by his counterpart in China. Also in early 2018, Baidu (China's leading search engine like Google) founder Robin Li commented in an interview that the Chinese people would trade their privacy for convenience. "If they [Chinese citizens] are able to exchange privacy for safety, convenience or efficiency, in many cases, they are willing to do that, then we can make more use of that data", he said.

Ironically, in China, Li's remark was not accepted by the public. The comment incited uproar among internet users, especially those upset with the search giant's invasive data collection practices. Chinese state media reported on the outrage users expressed online, citing comments like, "Who told you we are willing to give up our data"? As luck would have it, Baidu was sued in the same year by a consumer rights protection group in Jiangsu province for collecting user data without consent (the lawsuit was later withdrawn, after Baidu removed the function to monitor users' contacts and activities).

Indeed, the year 2018 could be hailed as the year when the Chinese public awakened to privacy. In the same year, Chinese user challenged another internet giant, Alibaba, among others, on personal data privacy. As mentioned in previous chapters, Ant Group, Alibaba's financial arm has launched Zhima (Sesame) Credit, an online credit scoring service that offers loans based on users' digital activities, transaction records, and social media presence. Users discovered that they had been enrolled in the credit scoring system by default and without consent. Under pressure, Alibaba apologized.

Increasingly, Chinese consumers are vocally standing up for their own privacy in front of internet giants in ways not seen before. At the same time, China is in the early stages of setting up a data protection regulatory system. Its 2018 *E-Commerce Law*, for example, has already incorporated data privacy protections similar to that of the General Data Protection Regulation (GDPR) of the European Union, whereas the Civil Code passed in May 2020 has enshrined the right to privacy and the principles of personal information protection for the first time. Contrary to Zuckerberg's characterization, China no longer provides him with a convenient counter-argument against privacy rules.

World's Largest Data Worker Pool in the Land of Big Data

To some extent, Mr Zuckerberg is correct. For AI research, the critical advantage China has is its data resources, as the density of people is proportional to the density of data. With more than 900 million of the population being internet users, plus the widespread and adoption of applications during the mobile economy boom, the data in China's consumer market has experienced surging growth. In this "mobile first" and "mobile only" market, people use their mobile phones heavily to shop consumer goods, order meal deliveries, buy movie tickets, and pay for almost all daily activities, which in turn, leave vast amount of data on the internet platforms.

Ture, for the AI race, China's great advantage lies in the Big Data it has. But its advantage is subtler than that. The background is that for all its advancements, AI is still very much as artificial as its moniker suggests. AI must be "trained" or "taught" for its cognitive abilities. An AI algorithm must digest vast amounts of tagged photos and videos before it realizes that a black cat and a white cat are both cats. Similarly, to learn to differentiate between cats and dogs, a computer must be shown pictures in which each animal is correctly labelled. For that training process (see Figure 8.1), information such as photos and videos must be gathered, and the details relating to the subject—for example, cats or dogs—needs to be labeled to tell the machines what they are seeing (in the future, the machines may become smart enough to teach themselves).

Therefore, data alone are not much use for developing AI. They must first be labeled. The boom in demand for data to train AI algorithms is feeding a new global industry—"data gathering and labeling". That's why data factory startups have sprung up in China, usually in relatively smaller, cheaper cities

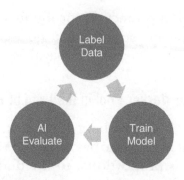

Figure 8.1   AI fed on labeled data

where both labor and office space are cheap. Many of the data factory workers are the kinds of people who once worked on assembly lines and construction sites in those big cities.

For labeling employees, the job is not super technical and could be boring. The work is usually about marking data points on photographs of people, surveillance footage, and streets. For example, labeling images of sleepy people that could be used by autonomous driving projects to identify drivers who are falling asleep at the wheel. Still, the new data job provides them the excitement to be the construction workers in the digital world, and the new industry brings some of the employment benefits of the tech sector to rural areas.

As a result, China has emerged as a key global hub for data collection and labeling, thanks to abundant data and cheap labor. If data in the new economy is (not entirely accurately) compared to oil in the old, the data collection and labeling businesses can be thought of as the crude oil refineries, turning raw data into the usable fuel for the AI industries. Just like the traditional "Made in China" jeans and T-shirt factories, the data factories are efficient and competitive in pricing, and they boast a robust mix of domestic and overseas clients as far afield as the United States, Mexico, Kenya, India, and Venezuela.

The AI race in face recognition is a great example. Underpinning Chinese unicorns like Megvii and SenseTime is a data supply chain through which data are collected, cleaned, and labeled before being processed into the machine-learning software that makes face recognition work. Many of the algorithms they used are not that different from others. China's data-labeling infrastructure, which is without peer, is an important factor for AI startups to stand out. Going forward, many traditional industries will also benefit from this machine learning supply chain as they undergo digital transformation.

As such, in 2020, China's central government issued an economic policy guideline to include "data" in "factors of production", joining the traditional elements of land, labor, capital, and technology (see Figure 8.2). China is at a pivotal stage of innovation-driven development, the government said, and data can have a multiplier effect on the efficiency of other factors. This initiative rhymes with the "new infrastructure" program announced at the May 2020 central government report, which has a laser focus on the data-related developments. Under this framework, personal data management and protection, unified industry data standards, public and private data sharing, among other key issues of the digital economy, are seeing accelerating development.

Meanwhile, the increasing importance of data has also resulted in the emergence of illegal data and fake data. The "black

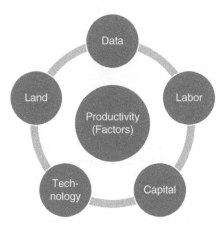

Figure 8.2   Data: The new addition to "production factors"

market" for illegal data is flourishing in China, driven by the large number and cheap price. The average number of data records involved in data breaches in China is multiple times higher than in other countries, and individual credit information and phone records can be bought for as little as US$.01, according to media reports. Just doing a simple search like "mobile phone data" or "personal data" on the social networks, one can easily find groups that are formed to sell and buy personal information.

The fake data industry is also flourishing, especially in the media and entertainment sectors, because user data means popularity and influence. On China's Twitter-like Weibo, the popularity of "internet celebrities" can be measured by the number of their followers, the times their posts have been liked, and their place on different ranking lists. Furthermore, because most of the entertainment platforms use data-driven AI algorithms to determine what contents to recommend to users, the value of data can be compounded. As the data-driven business grows, the fight against falsifications of data and statistics continues. (**See the "Water Army and Kris Who?" box.**)

Water Army and Kris Who?

In the sixteenth century, there were French "claques"—crowds paid to applaud theater performances. In the age of the internet, the "water army" does not even need humans. Nearly a third of China's internet traffic in 2018 was rated "abnormal", according to a report by third-party advertising data monitor Miaozhen Systems, and the resulting loss to advertisers reached more than 26 billion RMB (US$3.75 billion).

The internet water army is a group of fake accounts used by companies, celebrities, and influencers to inflate their social media followings (or criticize rivals)—so-called because they "pour water" into the online discourse. Just as in the e-commerce markets, vendors fake transactions (*shuaxiaoliang*) to inflate their trade volumes (**see the "Brushing the Sales Volume (*Shuaxiaoliang*)" box in Chapter 2**). In the digital marketing market, players can "brush the user traffic" (*shualiuliang*) to produce a viral effect.

According to media stories quoting one former employee of a big Chinese tech company, "Hiring water armies is part of every public relations manager's toolkit". Indeed, a simple search on any search engine or social network can find a large source of water army agencies. The professional agents offer a variety of services. For example, in what it called its basic package that costs about US$1.50 (10 RMB), a buyer can gain 400 followers (fake fans) on Weibo or 100 shares of a specific post. There is also a wide range of customized offers: the higher the quality of followers, the more active and realistic the account may seem to be—and the more expensive the service is.

Modern-day advertisers are dependent on market data to decide their brand ambassadors for online ads. However, this online data-driven system is vulnerable to exploitation by water army operators. The fake "influencers" have no real reach and fake fans don't buy anything. For foreign brands, their China office may choose brand ambassadors based on inflated metrics out of an eagerness to show the headquarters the continued "growth". In short, the fake user traffic has "muddied the waters" for brands.

The power of the Chinese water army has even reached foreign markets. On November 5, 2018, Kris Wu, a Chinese-Canadian rapper, managed to top Ariana Grande and Lady Gaga on iTunes top sales lists. For much of the day and into the night, the only non-Wu track in the top five was Ariana Grande's "Thank U, Next"—something Wu suggested was because of his loyal fans, not bots. But it was not clear how the Chinese fans were able to make their purchases on US platforms. When Kris Wu dominated the iTunes top 10 songs chart, many US pop music fans were left asking, "Kris Who?"

Of course, access to the most data in and of itself is not the most important factor in AI development. For example, after AlphaGo beat the best human chess player, Google further developed a more advanced version, AlphaGo Zero, which was not trained by historical data of games between humans at all. Instead, AlphaGo Zero was only taught the *Go* chess game rules before it started self-training by playing games against itself. Within a few days, AlphaGo Zero easily beat AlphaGo. Clearly, in addition to the quantity of data, other factors like new algorithms, computing power, and the kind of data available may be just as valuable.

Therefore, the analogy of data as the new oil is a bit flawed. The relationship between data and AI prowess is analogous to the relationship between labor and the economy. China may have an abundance of workers, but the quality, structure, and as the AlphaGo case illustrates, the governance (algorithm programming) of that labor force is just as important to economic development. Compared to China's vast data reserve, the United States holds advantages in data quality and data diversity. For example,

Facebook and China's WeChat both have more than one billion users on their social networks, but the Facebook users represent a far greater range of languages, ethnicities, cultures, and nationalities than those of the latter (which are mostly Chinese).

Nevertheless, before AI programs fully become self-supervised or unsupervised learning in which AI systems learn without human labeling, China's data reserves and large pool of data workers remain an important advantage, which feeds into the AI virtuous cycle—more data for testing, more AI products to implement, more data generated again from usage. That advantage is particularly valuable in the age of AI implementation (vis-à-vis discovery) as most engineers focus on running and testing AI algorithms on "Big Data" for faster speed, better predictions, less labor, less cost, and so on.

Furthermore, the Chinese labeling factories are upgrading themselves, as high-quality, large-scale, and integrated user data is currently under-supplied globally. As startups move into more advanced AI deep learning, it is the greater accuracy of labeled data that keeps a company thriving, rather than low prices and cheap labor. For example, data for autonomous cars and X-ray reading machines require specialist knowledge, and many data labeling players are setting themselves for more challenging tasks and higher profit margins. As China's Big Data industry expands, the public's data awareness rises.

From Human Flesh Search to Big Data Killing

In earlier years, Chinese users embraced the mobile internet with excitement, especially for the fun of online shopping and entertainment anytime and anywhere, at price much cheaper than before. CNNIC's 2014 annual report found that Chinese internet users' attitude toward the internet was mostly "trust and sharing". Their limited encounters with user data and privacy issues were those "human flesh search" incidents, but for most netizens, they were just problems of "bad people". **(See the "*Renrou-Sousuo* (Human Flesh Search)" box.)**

Renrou-Sousuo (Human Flesh Search)

Although one might suspect a link to internet pornography, the term "human flesh search" actually refers to Chinese internet users mobilizing themselves voluntarily to collectively search and post information online about specific people or public events.

The searches usually involve individuals suspected of crime (such as corruption), of foolish acts, or simply of conduct socially questionable to the millions of netizens. This usually takes the form of posting photos of people and accusations of misbehavior, but a massive number of internet users can oftentimes collaborate to very quickly reveal a lot of information on an offending person.

The "bodily" aspect of this search term probably originated with a 2011 event involving a model's identity. In 2011, someone posted a photo of a pretty young woman on Mop.com – a discussion website popular with teens and those in their early twenties, claiming that the photo was of his girlfriend. Suspicious visitors at Mop.com soon discovered that the beauty was a model and collectively posted her information as proof that the original poster was lying.

In this case, internet users penetrated beyond a person's facade to reveal that person's "flesh", so to speak. "Human flesh search" thus stands for performing a deep search on the internet. Another theory is that the name refers to the broad knowledge scattered across social networks and the collaboration involved in searching. In this context, "human flesh" stands for "people in general". With the amount of information that can be exposed on the internet, human flesh searches are very powerful, sometimes disruptive, and very often controversial.

In some sense, the best example of "human flesh search" was a dressing room sex video in July 2015. A young couple filmed themselves with a smartphone when they had sex inside the fitting room of a Uniqlo clothing store in Beijing's entertainment district Sanlitun. The video went viral on Chinese social media after it was posted online. It did not take long for the Renrou-Sousuo web users to dig up the porn enthusiasts' social network accounts and their other details (although the couple denied that it was them featured in the video), which probably helped Beijing police arrest at least four people for their involvement in the sex tape.

But things quickly changed by the emergence of the large internet platforms. As is mentioned in previous chapters, many internet giants have expanded their businesses into "super apps". For example, Tencent openly vowed to become the fundamental platform for China's internet: a platform equivalent to the "water and electricity supplies in daily life".

They are active in various areas, from food ordering to taxi hailing, e-commerce to money lending, and they collect and possess massive user data from their numerous activities. Consequently, these large platforms are able to profile their uses in striking details.

As the "mosaic theory" suggests, disparate items of information, though individually of limited or no utility to the owner, can take on added significance when combined with other items of information. In cyberspace, there is a lot of different information that a user would never think would be able to identify a person. But when a computer combines the different pieces, it can see connections in ways that humans cannot. When a digital platform combines different sets of data, either from different service lines of the same platform (e.g. Meituan's food ordering and bike sharing services) or from third-party data vendors in different sectors, the power of user data integration grows exponentially.

On the one hand, the platforms have used Big Data analysis to provide users with more personalized services and faster services. On the other hand, the data power of many platforms has also aroused public concerns that Big Data could be abused. They quickly learned the importance of data and privacy through "Big Data killing"—the more personal data the platforms have, the more users have to pay (**See the "*Dashuju Shashu* (Big Data Killing)" box.**)

Dashuju Shashu (Big Data Killing)

Shashu literally means "killing someone that a person is acquainted with", and it is a term coming out of China's market economy, which refers to the situation where a person takes advantage of another who innocently believes the person is a friend and acts in his or her interest.

In the data economy, it evolves into "Big Data killing", as the internet platforms capitalize on their regular users when their spending habits are well known by the platforms. The "killing" refers to the situation where, for the same goods and services, the price shown to old customers is more expensive than that to new users. In economic terms, "Big Data killing" is a form of price discrimination.

Big Data killing illustrates the power and value of data. Because the internet platform has no knowledge about the new user, it would offer a relatively low product (or service) price so that the new user can enjoy the "sweetness" of the first experience (at the same time, the platform gathers his or her personal data through the user's platform registration and related transaction). Meanwhile, the internet platform offers relatively higher price to existing users, especially those who are analyzed to have high spending power and low sensitivity to pricing.

For example, online travel agency (OTA) sites are where Big Data killing is prevalent. Many users have discovered that when they try to book air tickets or hotel rooms, the price would be higher for a frequent user of the website than a newcomer. Online car-hailing platforms are also found to offer different prices in the same region to different users. Similar phenomenon is also reported to occur in online shopping, online ticket purchases, video websites, and many more fields.

In April 2018, iMedia Research released "2018 China Big Data Killing and User Behavior Report." According to its analysis, 77.8% of surveyed internet users believed that service applications using Big Data for differential pricing were unacceptable; 73.9% did not know that internet applications were using Big Data to categorize different user behaviors. The report suggests that a high percentage of internet users are unaware of such a practice. Many of them are insensitive to prices and are likely to become the target of Big Data killing.

The regulators took action as the issue gained social traction. Big Data killing is prohibited by China's E-Commerce Law that became effective January 1, 2019. The law provides that when e-commerce operators provide a search result for a good or service, it should simultaneously provide the consumer with an option to see results that do not target his or her identifiable traits. In other words, merchants could offer customized services and products to users, but users also have the option to see general offerings that are not based on personal data.

As the Big Data killing illustrates, the notion that Chinese internet users care little about giving up personal data is not true. The more direct reason is that there is still a general lack of understanding as to how data is collected, categorized, and used by social platforms. Nevertheless, after years of Chinese internet companies building business models around Chinese people's lack of awareness about privacy, users are becoming more knowledgeable, and they are becoming angry with companies abusing their personal information.

The AI Age: Privacy, Security, and Value

In the age of AI, internet users have more direct dealings with data issues. Because data has become a critical resource in AI and data-driven technologies, the internet giants are more often *proactively* collecting user data. Furthermore, they are collecting every aspect of user data, whether identity data, network data, and behavioral data (see Figure 8.3). Take precision marketing, for example. Users' data are analyzed and based on the different characteristic labels they are given (e.g. "makeup lover", "sports fan", and "keen to travel"). Then, companies show specific advertising messages to potential customers based on the matching of labels.

- **Identity data**: includes basic information about a person, such as name, gender, mobile phone number, and identity card number, which are mainly used to authenticate users' identities. With China's real-name registration system, when users enter almost any app for the first time, they are required to register with a phone number as phone number can be traced to their national ID.
- **Network data**: contains location data, log data, and device information. For example, Alibaba's mobile payment service Alipay has incorporated social functions

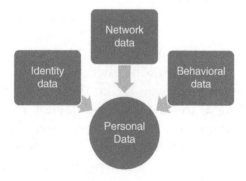

Figure 8.3   Personal data—key resource for AI

through which it encourages users to share location data as well as personal information and purchasing habits with others.

- **Behavior data**: When users browse websites or Apps, those behaviors are recorded to extract user behavioral habits. For example, from the patterns of Facebook Likes, data analysts could predict the users' sexual orientation, religion, alcohol and drug use, relationship status, age, gender, race, political views, and more.

In June 2018, the State broadcaster China Central Television (CCTV) and Tencent Research surveyed 8,000 respondents on their attitudes toward AI as part of CCTV's China Economic Life Survey. The results show that 76.3% of respondents see AI as a threat to their privacy, which means three out of four people in China are worried about the threat that AI poses to their privacy. Average internet users are increasingly aware of the serious personal data issues involved—privacy infringement, security risk, and commercial value (see Figure 8.4).

First, the over-collection of user data. Because the legal development lags the age of Big Data, network service providers globally have all taken liberties with the collection and use of personal information in mobile apps. The issue is most prevalent in China due to its unrivaled "mobile only" user

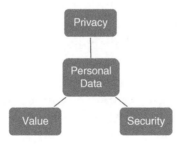

Figure 8.4   Serious issues around personal data

population. In 2018, China accounted for nearly half of global app downloads according to a report from AppAnnie.

In 2018, the China Consumer Association, the national organization responsible for monitoring goods and services and protecting consumer rights, evaluated 100 Chinese mobile apps of a wide range of service sectors, including social messaging, online shopping, mobile payment, transportation, news, and photo editing. The survey found that many apps were collecting user data that's not relevant to their primary function, or they failed to inform users when collecting and using such data.

For example, a photo-editing app should have no right to demand user location, ID number, or fingerprints. And why does a weather forecasting app ask for the access to user's contact list? In the report, the association concluded that 91 of the 100 apps over-collected user data, such as location, contact lists, and phone numbers. The list included some of the most widely used apps in China, such as the messaging app WeChat and the microblogging platform Weibo. Furthermore, the ride-hailing service Didi Chuxing, the popular social e-commerce app Pinduoduo, and the short video livestreaming app Kuaishou received the survey's lowest possible consumer privacy rating.

Second, with more data gathered, there is more data leak, theft, and trade. According to the Internet Society of China, data leaks affected over 80% of netizens in 2017. In another 2018 survey, but by the China Consumer Association, 85% of respondents said that their data had been leaked, including phone numbers sold illegally or bank account information hacked.

Numerous data theft cases have made headlines of late, and the case of Apple employees in China stealing iPhone user data for sale and the Momo user data sale probably were the most high profile. In December 2018, the account credentials of 30 million users of the dating app Momo were found to be stolen and sold online (Momo app lets users find each other via their profiles and shared locations and is often called "China's Tinder"). Based on a series of screenshots from the Chinese dark web, the user data—including phone numbers, username, and passwords—were reportedly sold for as little as 200 RMB (around US$30).

Buying Faces

In rural areas, such as Henan province in the middle of the country's heartland, the peasants of small villages found that they could trade images of their faces for daily supplies. Villagers were asked to stand in front of a camera and slowly rotate side to side to have their face pictures taken. These face pictures would be used in AI software to distinguish between real facial features and still images. In return, the peasants were given kettles, pots, and teacups.

"Buying faces" actually is not a China-only phenomenon. In the summer of 2019, Google announced a face unlock feature for its upcoming Pixel 4 phone, which claims to be just as accurate and fast as the iPhone's Face ID. In various cities across the country, Google employees were reportedly doling out US$5 Starbucks and Amazon gift cards to people on the street in exchange for a facial scan. Google confirmed that such "field research" was an effort to collect diverse face-scanning data to ultimately improve the accuracy of its upcoming Pixel 4's facial recognition technology.

Third, the commercial value of data. As mentioned, China has emerged as a key hub for data collection and labeling thanks to abundant data and cheap labor. The "data gathering and labeling" factories actively seek ways to gather new data, which includes heading to the street to talk to individuals and directly purchase personal data. (**See the "Buying Faces" box.**) Therefore, even ordinary users in the remote areas are educated about the intrinsic value of data.

More experienced users are now fully aware of the value of data in the digital economy, as they see every day the internet giants announcing new data-driven strategies or some companies experiencing significant valuation swings from their data-related practices. For example, the selfie beautification app Meitu, in December 2018, saw its share price plummet 10%

after the China Consumer Association published the report criticizing it "over-collecting" personal data.

The increased "data awareness" is best illustrated by the Zao app case in September 2019, when China had its own version of the FaceApp privacy storm like in the West. Using artificial intelligence and machine-learning techniques, the Zao app allowed users to swap faces with celebrities in movies or TV shows. It went viral as a tool for creating deep fakes, but concerns soon arose as people noticed that Zao's user agreement gave the app the global rights to use any image or video created on the platform for free and sell them to third parties without users' consent.

After a public outcry regarding these controversial user privacy terms and questions over data safety, the company clarified that the app would not store any users' facial information. Under pressure, the company also declared that once a user uninstalls Zao or deletes their account, the app would remove related information as required by regulations and "ensure the safety of personal information and data in every possible way".

New Rules for a New Game

Clearly, in China the era of free data collection—"freely", "free of responsibilities", and "for free"—is over. This tidal shift in public opinions is not that striking in the context of China's economic developments in recent decades. A decade ago, for example, few people in China were worried about air pollution or the related public health risks. Fast forward to today, air pollution is among Chinese people's foremost concerns, and local governments are working on more "blue sky days" in the same urgency as more GDP growth. As the country continues to digitize its economy, privacy infringements and data breaches have become hot-button issues in China.

Until recently, China's data privacy framework consisted only of a patchwork of fragmented rules found in various laws, measures, and sector-specific regulations. The *Cybersecurity Law*, which came into effect on June 1, 2017, included for the first time a set of data protection provisions in the form of national-level legislation. Then the 2018 *E-Commerce Law* incorporated important concepts from the General Data Protection Regulation (GDPR) of the European Union. In 2019, a series of administrative regulations were issued to set the personal data protection standards.

In May 2020, China's national legislature approved the nation's first *Civil Code*, enshrining the right to privacy and the principles of personal information protection for the first time. Around the same time, the work report from the Standing Committee of the National People's Congress mentioned that the much-watched *Personal Information Protection Law* and *Data Security Law* are currently in the process of formulation. These laws and regulations will collectively form a comprehensive framework for individual data rights and protection.

Underpinning the Cyber Security Law is the Chinese concept of cybersovereignty. Relating to individual data rights, it imposes data privacy obligations on network operators, which is defined as any system that consists of computers or other information terminals, and related equipment for collecting, storing, transmitting, exchanging, and processing information. The law regulates how organizations should protect digital information, and outlines measures to safeguard internet systems, products, and services against cyberattacks. The law also requires that operators of key information infrastructure store important business and personal data "locally" that they collect from their operations in China. In compliance, Apple, Inc. has built a data center in Guizhou (the Big Data city introduced in Chapter 7) with Chinese partners.

China's E-Commerce Law, which was passed in August 2018, took effect on January 1, 2019. This law aims at improving regulation of the flourishing online market, specifying various regulations concerning operators, contracts, dispute settlement, and liabilities involved in e-commerce; meanwhile, with regard to the increasingly complex online market operations, it also addresses important aspects of consumer data protection and cybersecurity, such as:

1. **No price discrimination based on individual user data**. When providing the results of searches for commodities or services for a consumer based on hobby, consumption habit, or any other traits thereof, the e-commerce business shall provide the consumer with options not targeting his or her identifiable traits.

2. **No automatic tie-in sales**. If an e-commerce business performs tie-in sale of commodities or services based on consumer profile data, the platform shall request consumers to pay attention in a conspicuous manner, and it shall not set the said tie-in sale as a default option.

3. **Protection of personal data**. An e-commerce business shall expressly state the means of and procedures for search, correction, or deletion of user information; when receiving an application for search, correction, or deletion of user information, an e-commerce business shall, upon verification of identity, permit search, correction, or deletion of user information in a timely manner ("the right to be forgotten").

While the Personal Data Protection Law was still being drafted, the Cyberspace Administration of China (CAC), the highest administrative regulator of the Internet, in May 2019, issued *Measures on the Administration of Data Security*. These measures lay out specific rules regarding the dos and don'ts for how internet companies collect and use customer data, and it is the most direct, comprehensive, and highest authority regulation in China on individual data protection.

The CAC Measures are built on the earlier "Personal Information Security Specification" set by the National Information Security Standardization Technical Committee (TC260), a separate government body tasked with advising central government agencies, including CAC. In 2018, the committee created the "Personal Information Security Specification" as the best practice guide for the collection, processing, and sharing of sensitive information by digital businesses.

The specification was further amended in early 2019 to add in stronger provisions to refine existing requirements, especially those relating to personalized content, user consent, and third-party controls. For companies to act on user data, users must provide informed consent for all acts of sharing and processing, and have full rights to their data, including viewing, correction, and deletion. Furthermore, companies must have valid grounds for collecting personal data, and must present a transparent privacy policy clarifying why and how this data will be used. And they must collect only the minimum amount of data required, cannot use it for ancillary purposes without consent, and can only retain it—while encrypted—for the minimum time necessary.

Essentially, the CAC Measures have turned the "Personal Information Security Specification", a nonbinding policy suggestion that had no applicable penalties for breach, into the first binding administrative regulation in China on data. Before China's *Personal Information Protection Law* is fully enacted, the CAC Measures effectively set the personal data protection standards in China, and for the market, it provides a reference for the direction of the upcoming national law.

For example, the CAC Measures focus, in particular, on how users can get more control of their data in mobile apps. That's quite understandable because nearly 99% of China's 900 million internet users access the web via mobile phones. It sets out the following as situations that involve illegal or excessive collection of user information by mobile apps:

- no publicly available user data rules;
- no explicit statement of the purpose, method, or scope of collecting user information;
- information collection without consent;
- collecting personal information unrelated to the service provided;
- failure to delete or correct personal information as required by law; and
- bundle main service with additional functions that force the user to provide personal data for all services.

In parallel, the CAC, together with the Ministry of Public Security, the Ministry of Industry and Information Technology, and the State Administration for Market Regulation, have launched a national campaign to inspect smartphone apps to determine if they illegally or excessively collect users' information. As a result, by July 2019, a group of widely used apps had been ordered to correct their data collection practices, and 10 apps, including from the Bank of China, were found to have no user privacy rules. Another 40 apps, including those from many online financial platforms, were highlighted for "serious issues" in their data collection. Just days before New Year 2020, these four administrative agencies jointly published a new rule designed to provide a practical standard for identifying illegal collection and use of personal data by app developers.

These are all major steps forward to develop more protections for personal data, at a time when individual users have little bargaining power against popular apps and large platforms. However, the personal data issues are becoming increasingly complex, as more aspects of everyday life become a digital experience and more stakeholders are involved in the data debates. As a result, there are still more questions than answers, best illustrated by the Didi Chuxing case in the following section.

Ride-hailing Sex Incidents: More Questions than Answers

The year 2018 started as a bright one for Didi Chuxing. It had outlasted several domestic competitors, acquired Uber's China business, and become the unquestionable leader in China's ride-hailing services. It also joined two tech stars, Toutiao/Bytedance and Meituan, to form "TMD", an acronym for the three high-valued unicorns that are poised to be the next generation tech leaders. But soon, Didi was confronted with its biggest crisis ever since it was founded six years ago. The turning point was two murder cases out of its once popular "Hitch" (carpooling) service, which allowed ordinarily car owners to pick up passengers going in the same direction.

In May 2018, a female flight attendant, aged 21, went missing after she took a Didi hitch-hiking ride in Zhengzhou, in central China. Later, the police discovered her body and found that she was raped and killed by her driver, which led to an angry public demanding urgent action on safety. Just three months later, another 20-year-old girl was raped and killed in similar circumstances in Hangzhou, and the angry voices—of users, regulators, academia, and police—rose to a crescendo.

The two tragic events cost Didi enormously. Didi suspended Hitch (which was estimated to provide 10% of Didi's bookings) and dismissed two of the executives who led the service. According to market analysis, Didi's brand value dropped significantly soon after these killings and it never recovered to reach its May 2018 peak. After more than a year, in January 2020, Didi restarted the Hitch service in Shanghai and a few more cities. Still, Didi's name remained tied to sexual incidents. When a driver who claimed to be working for Didi posted a film of his rape of a female passenger on a live-streaming platform in June 2020, Didi spent a moment in shock. **(See the "Drugged Rape Live Broadcast" box.)**

Drugged Rape Live Broadcast

Live streaming has prompted creativity in China. On June 11, 2020, some netizens stumbled on a live broadcast room titled "Late Night Didi Trap" on an underground live broadcast app called "star link" (*xinglian*). The male anchor is named "outdoor car shock", who claimed to be a Didi ride-hailing driver in Zhengzhou, Henan province.

The "Didi driver" announced that he was going to find a female passenger as a "prey" for sex in the car. He boasted his "success story" of the previous night, showing himself as an "old driver" (a pun for "veteran"). At 2 a.m., the driver saw a young girl waiting for a car in front of a nightclub. He looked out of the window and asked the victim loudly if she needed a ride. Then the online audiences watched in horror—a 21-minute video—as the driver drugged her using a spray before sexually assaulting her in the car.

Hours later, some people posted the link in social networks for the public and law enforcement agencies' attention. Apparently, it was not that easy to confirm the identity of the driver, as on that evening, Didi Chuxing released a statement at its official Weibo account: "We are aware of the 'Didi driver sexual assault live broadcast'. We have called the police and are urgently verifying it. We will report it to the public timely".

Although the incident seemed to have been staged, as medical professionals commented on social platforms that there is no known drug that can knock a person unconscious for 10 minutes with just a few sprays, the Didi management probably held their breath until Zhengzhou local police released the result of the investigation. According to the police report, the incident was a young couple's stunt for live streaming. Aged 22 and 21, respectively, the adventurous husband and wife designed the plot of a car-hailing driver raping a female passenger. The show did attract paying audiences, but ended badly: the police arrested both for disturbing public order.

Because of the Didi incidents, the public started a discussion on improving the data management of mobile apps for public safety, and soon they realize the underlying issues are much more complicated than they initially thought:

First, the drivers' identity data. In the first murder case, the killer Liu's original driver registration was rejected by the Didi system based on his background information. He then used his father's ID to register, and his father also helped him to pass the facial recognition test. After that murder, Didi tightened up its registration process as well as its facial recognition requirements.

In the latter case, the second driver had passed background checks and had no prior criminal record. He had registered with the company using his authentic ID, driving license, and vehicle registration certificate, according to Didi announcements. However, the media *Beijing News* found out that the driver was seriously in debt, having taken out loans from more than 20 online lending platforms in the month before the murder. The report suggested that Didi should have gathered more information on the driver and found the warning signals. But how much driver data is enough for Didi's duty of care? To what extent will stronger screening checks become "excessive collection" of drivers' personal data?

Second, the passengers' identity data. Before hitching drivers accepted someone's ride demand, they were able to see information of the potential passenger, which used to include the person's profile picture, gender, and reviews by drivers from earlier rides. In the murder cases, the drivers used such information to select the victims. After the murders, the passenger's personal information was no longer on display for the drivers, and the reviews were limited to tags relating to the rides, such as "punctual" and "polite". This approach probably could have prevented the two driver killers picking on the young female passengers as their rape-and-murder victims, as Didi drivers reportedly had been reviewing female passengers based on their appearance, including tags like "pretty", "good body", and even in some cases "sexy".

On the other hand, for the safety of the driver, should the driver have access to some passenger information before taking a ride with a stranger? Just two months after the two instances of drivers killing passengers, in December 2018, a Didi driver was killed by a passenger during a midnight trip in southern China Guizhou province. He was robbed of cash (and his digital wallet) before being murdered, and his body was thrown from a bridge (there had been similar cases before that). Surely, Didi was also facing pressure from drivers' families—how much passenger information should be displayed for the drivers?

Third, the recording of the rides. After the two murders, Didi has ramped up safety features, such as adding an emergency button in the app. Furthermore, all conversations during Didi rides are recorded. To protect the passengers, the recordings were not kept on drivers' smart phones; instead, they were uploaded to company servers with encryption, which are kept for seven days. In July 2019, the company said that since September 2018, 20% of the rides even have video recordings.

For the audio and video recordings, the public voiced concerns of privacy, both from the perspective of the passengers and drivers. What if the passenger was using the ride time to dial in a conference call for the company he or she works in?

In such a case, Didi may pick on proprietary information and business secrets. And what if the driver was bored by the long ride and wanted to talk intimately with his girlfriend? Should such personal and colorful conversation be recorded and uploaded to Didi's servers?

Fourth, who owns the recorded data? In 2018, Didi resisted turning over data to law enforcement authorities after two passengers of the app were murdered. The company cited privacy as a justification, and it also pointed to the vague definition of *data ownership* in fragmented regulations.

The operation of ride hailing, under the current legal system, is regulated by the Ministry of Transportation, which didn't define a requirement for companies like Didi to submit operation data. The Ministry of Public Security, on the other hand, does not run a government supervisory platform that monitors the data of mobile services like Didi. In the murder cases, Didi was much like a hotel that experiences a murder case and does not necessarily have to submit its hotel operation data to the police. In short, even a fundamental question like "Who owns the data" has no clear answer.

Overall, the Didi incidents have illustrated the complexities of the data-driven economy. Issues around data can involve balancing multiple legitimate values at once—any combination of privacy, national security, criminal justice, civil liberties, innovation, freedom of expression, consumer protection, anti-trust, and national economic interests can be at stake in any single situation. With the spread of the 5G network and data technologies—faster data rates, massive device connectivity, more powerful data operation capabilities—the conflicts among the rights and responsibilities of various stakeholders will only be further exacerbated.

Civil Code 2020 and More to Come

China's new *Civil Code*, which will take effect at the beginning of 2021, marks a key step forward in developing the legal framework governing individual data privacy. Among a sweeping package of numerous civil laws, for the first time, privacy is given statutory force as a personality right. The code devotes one entire chapter to addressing personality rights, which covers people's rights to name, title, portrait, reputation, and privacy (i.e. privacy is on a par with other fundamental rights) while adding new articles on protecting personal information. Far from a complete data law, the Civil Code nonetheless has established the fundamental principles of personal information protection, paving the way for future legislation such as the awaited *Personal Information Protection Law* and *Data Security Law*.

Compared to the Cybersecurity Law effective since 2017, which was the first piece of national legislation to protect online personal information, the Civil Code covers privacy and personal information in a more "civil right" context and in more details (see Table 8.1). To counter the power of digital

Table 8.1   Selective improvements in 2020 Civil Code

	2020 Civil Code versus 2017 Cybersecurity Law
Scope of personal information	Expands the definition of *personal information* to include, among others, personal e-mail address, location/ traveling information, and biometrics information
Data management	For data-driven businesses, introduces the concept of "processing data" to expand the concept of "using data", which includes collecting, storing, using, adapting, transmitting, providing, and publicizing personal information
Protection of minors and People with limited civil capacity	Emphasizes the need to obtain the consent of a guardian for the collection of personal information of persons without civil capacity, such as minors, or persons with limited civil capacity
Obligations for digital platforms	Introduces a number of obligations for information processors (tech companies), for example, personal information cannot be illegally provided to others without the consent of the person; technical and other necessary measures must be adopted to ensure the security of the personal information they collect and store to prevent information leakage, tampering, and loss
Users' access to own data	Clarifies that individuals have the right to access and get a copy of their data, which was not clearly stated under the Cybersecurity Law

platforms, the code introduces the concept of personal information processing (similar to the definition of *processing* in GDPR), which includes collecting, storing, using, adapting, transmitting, providing, and publicizing personal information. Such changes will have significant impact on Chinese mobile apps that use individuals' personal behavior data and location data to promote lifestyle services like restaurants and food deliveries near the users' neighborhoods.

Of course, there are also compromises in the code's efforts to protect privacy. For example, although the code protects personal information under personal rights, it does not recognize personal information as an inherent right for everyone. This implies that an individual will only have an economic interest in his or her personal information rather than a personality right, which probably reflects resistance from the e-commerce industry. In some provisions, there is a lack of details: for example, how long data collectors can keep people's information and under what circumstance they must delete data.

And many areas seem to be left intentionally for future legislation. Take the "consent" of users, for example. Now that biometric information that powers facial recognition system is included in the definition of personal information, the consent mechanism becomes more complicated. On the one hand, when facial recognition is used for *confirmation*, such as unlocking one's iPhone—ie, identifying yourself to your own device—it's easy to design the consent mechanism. On the other hand, when it is for *identification* for public security purposes, it is hard to send notices and obtain consent from consumers in situations where facial data needs to be captured from large numbers of people in the public sphere.

Furthermore, the ongoing coronavirus pandemic has continuously created novel data and privacy controversies. For epidemic control and prevention efforts, a vast amount of individual information is collected by governments to keep close tabs on the health and location data of local populations. Under ordinary circumstances, sensitive patient-linked medical records can and should be kept private; but during the extraordinary crisis, many governments are willing to overlook privacy implications in an effort to save lives. Because governments' efforts are sometimes hosted by internet giants to take advantage of their existing user networks, the private companies are taking on public mandates, which empowers them to collect and manage more individual data than before.

Many complex questions like those previously mentioned are expected to be answered in more detail by the forthcoming *Personal Information Protection Law* and *Data Security Law*, which will pose a tough challenge for Its drafters to balance the considerations of individuals (personal privacy), enterprises (business development), and governments (national and public security). Since they are already in the process of formulation at the National People's Congress, they may become effective as early as 2021. These two specialized laws will be a major step toward regularizing this patchwork affair of personal information protection into an integrated, comprehensive framework (see Table 8.2).

Meanwhile, Chinese tech startups are also active in innovation of digital privacy solutions. "Federated Learning", for example, is used by AI companies to build machine learning models across multiple data owners while ensuring that no private data is leaked, and data is used in full knowledge of the multiple data owners. Blockchain ventures are using DLT (distributed ledger technology) to facilitate data flow and protect data

Table 8.2   China's data law framework

	Laws governing Data Privacy and Security
2017	Cybersecurity Law
2018	E-Commerce Law
2019	CAC Measures on the Administration of Data Security
2020	Civil Code
2021 (expected)	Personal Information Protection Law; Data Security Law

privacy simultaneously. These are important tech startups because the issue of privacy protection can only be solved via technical experts and policy makers going hand in hand.

LeakZero, for example, is a search engine that does not track users. Previously the chief security officer at Amazon China and Xiaomi, Yang Geng the founder is passionate about building technology to protect user privacy. The LeakZero users can use a random name generator to create an ID and then surf the web anonymously. Then, when they communicate on a messaging app such as WeChat, LeakZero is "layered" on top to encrypt their messages into nonsense string of letters and numbers. (The recipient must also use LeakZero's app to decrypt the message.) By design, LeakZero company cannot find out the names of the app's users or even know how many there are; however, according to 2019 media reports, the encrypted search and messaging app has more than 34 000 aliases.

As the largest mobile internet user market, China's individual data rights framework (and "privacy tech" innovation) has profound global implications. Because of foreign internet giants operating online in China and Chinese companies operating globally, China's cyberspace law and policy is relevant worldwide. Although many foreign critics find it hard to understand that China's increased privacy protections for consumers coincides with its embrace of high-tech government surveillance, China's legal development as part of the global digital governance system simply cannot be ignored.

China's data law framework may stimulate the United States, which still has no national-level position on data protection, to expedite relevant legal measures. (In February 2020, Mark Zuckerberg published an opinion piece in the *Financial Times* newspaper titled "We Need More Regulation of Big Tech". The article, however, mostly focused on the "transparency" and "openness" of major tech companies, instead of data privacy issues. Therefore, it is not clear whether Chinese data laws and regulations have influenced his thinking.)

For Europe, GDPR is not explicitly tied to more far-reaching goals for national security and social stability. However, the European Union in February 2020 unveiled a plan to restore what officials called "technological sovereignty", which aims to boost its digital economy and avoid its over reliance on non-European companies. As such, new laws to reflect more "data sovereignty" can be expected in the European Union. Furthermore, China's framework may also provide a reference point for major emerging economies such as India, Brazil, and the ASEAN countries when they look to regulate cyberspace activities and emerging technologies.

Overall, it's good to see more actions on individual data protection emerging globally. These may lead to a global dialogue and the formation of a global practice on this critical issue, avoiding a potentially fractured global legal landscape. As Chapter 9 shows, cross-border data acquisition and flow is the frontline battlefield of today's geopolitics, which may derail the momentum of global digital economy. That's why the concept of "Digital Silk Road"—global connectivity and mutual development, discussed in Chapter 10—is such an important initiative for all stakeholders.

PART

III

Shared Digital Future

China in Asia and the United States in the West are forming two leading innovation centers of different strengths. China's digital economy is expanding overseas, both forming development partnerships with emerging markets and challenging America's tech supremacy.

C H A P T E R

The Tech Cold War

- ZTE's Shot Blocked
- TID—Not All National Security Is the Same
- Critical Technology and Semiconductor Chips
- From Hard Tech to Invisible Data
- China and the United States—Mutually Self-Sufficient?
- Europe, Japan, and More—the Divided Tech World
- No Winner in Balkanization of Cyberspace

ZTE's Shot Blocked

When US President Trump signed an executive order in early 2018 prohibiting the Chinese telecom firm ZTE from purchasing US electronic components for breaking United Nations sanctions and doing business with North Korea and Iran, many Americans heard about ZTE for the first time. Some may have seen ZTE-branded smartphones before in the US market, but very few would have known that ZTE was a pillar of China's high-technology sector. Across the Pacific, Chinese people watched incredulously as ZTE's operation came to a screeching halt within three weeks, exposing the soft underbelly of

China's technological ambitions—the reliance on US semicon-ductor microchips—for the world.

Although ZTE was a major supplier of telecom gear used by carriers around the world, its name was still unknown to many people outside of China. ZTE was founded in 1985 at the tech metropolis of Shenzhen in Southern China. Traditionally, its main business was supplying telecom equipment and network services to global telecommunications carriers. In recent years, the consumer-device business that included smartphones was a new growth area for the company. Among Chinese brands, ZTE was the only Chinese manufacturer with a sizable mar-ket share in the United States, reaching a top three Android phone-maker ranking.

Since the end of 2014, it had become the fourth largest smartphone brand in the United States – only behind Apple Inc., Samsung Electronics, and LG. For the US market, ZTE focused on young sports fans. For these fans, the most important feature was the camera, to catch fast-moving objects and scenes. According to ZTE, the front camera had "incredibly fast auto-focus" (in ZTE's terms), and it would allow users to take self-ies simply by smiling. Moreover, ZTE partnered with National Basketball Association (NBA) teams to increase its brand aware-ness in the US market and worldwide. (**See the "ZTE's Winning Picks" box.**)

ZTE's Winning Picks

During the 2014–2015 NBA season, ZTE selected three teams to start its sports-themed marketing campaign: the Golden State Warriors in San Francisco, the Rockets in Houston, and the Knicks in New York City. San Fran and NYC are the two cities in the United States with the longest history of Chinese immigration and the largest population of ethnic Chinese residents. The Houston Rockets has developed a big Chinese fan base since the 7-feet-3-inch center Yao Ming was acquired from China, and it became the most broadcasted team in China. (Years later, the America-born-Chinese Jeremy Lin, "Linsanity", also played at the Houston Rockets after a stint with the New York Knicks.)

The Golden State Warriors and Houston Rockets were among the top winning teams for the season, and both went to the Western Conference Finals. Hence, that series was already a big win to the ZTE brand. After eliminating the Rockets, the Golden State Warriors went to the finals and won the NBA Championship. It was a huge win for the franchise that had not won a championship for 40 years, bringing priceless marketing presence for ZTE.

Under the partnership deal, the ZTE brand was promoted around arenas, on the courts, and via giant LED displays outside of the arenas. The players were not necessarily required to use ZTE phones, but they got one if they asked for it. For example, when Jeremy Lin was with the Houston Rockets, he took a selfie with a ZTE phone, which led to a YouTube video "How to Take the Perfect Selfie with Jeremy Lin"!

The New York Knicks did not have a winning record, but the team had the ultimate all-star of all-stars: New York City. Every day, there were more than one million people passing by the ZTE LED sign outside Madison Square Gardens, the arena for Knicks' games. Later, in the 2015–2016 season, ZTE further signed new deals with two more leading NBA teams—the Chicago Bulls and Cleveland Cavaliers, making its basketball sponsorships total five major NBA teams.

It is worth noting that at the ceremony to add the Knicks to its list of team partners, ZTE's North America operation CEO Lixin Cheng was awarded a No. 8 jersey to celebrate the deal because in Chinese society the number eight is the luckiest number of all. The good luck—probably as important as ZTE's technical superiority and smart marketing—was a bit short-lived, as the company's expansion into the United States was brought to a grinding halt by the Trump administration's full-court press since 2018.

ZTE smartphone's leading position in the US market, one of the largest and most competitive markets in the world, is highly significant. Not that long ago, China's tech industry was known primarily for low-cost, cheap knock-offs and copied internet business models. Fast forward to today and Chinese companies are moving to the forefront of global technology innovation, particularly when it comes to hardware.

In fact, when President Xi Jinping visited Tajikistan during the Shanghai Cooperation Organization (SCO) Summit in September 2014, the ZTE smartphone was on his list of national gifts. The specific ZTE model used for national gifts was reputed as one of the world's most secure handsets, thanks to ZTE's recruitment of Blackberry's research team. (**See the "ZTE Picks Up Blackberry" box.**)

ZTE Picks Up Blackberry

Research In Motion (RIM) was once the pioneer of the smartphone segment with its BlackBerry device, and for many years, the company was a gem of Canada. But the market share of Blackberry in recent years was quickly eaten up by Apple and Android phones. In 2013, the company experienced serious financial distress and went into restructuring.

During the restructuring process, RIM announced multiple rounds of lay-offs, and many tech companies, including ZTE, Huawei, Samsung, and TCL all paid close attention to the departing research team. In March 2014, RIM announced a round of layoffs, including the senior engineers for chip design and product design from its Waterloo headquarters, and the Blackberry products research center in Ottawa.

ZTE was so determined to expand in North America, and to nail those talents, that it created a special department inside its human resources department dedicated to that recruitment task. In the end, ZTE beat the competitors and succeeded in cherry-picking more than 20 engineers from Blackberry, giving its North America R&D capabilities a substantial boost.

The most famous know-how of RIM is communication security, which explains why blackberry devices and services have been widely used by corporations, and even the Pentagon and NATO. In the hyper-connected mobile internet era, the security issue is more important than ever, and ZTE (Canada) released a new smartphone model reputed to be one of the world's most secure handsets. As is noted in later sections of this chapter, if the Blackberry recruitment had happened more recently, it likely would have been blocked by the US CFIUS (Committee of Foreign Investment in the US) for national security concerns through its long-arm jurisdiction.

Historically, China has mostly used traditional artwork, such as silk, porcelain, and paintings, as national gifts in diplomatic contexts. Using domestic smartphones as a "national gift" showed the progress China had made in the technology sector in recent years, and it also demonstrated China's determination to transform itself from the "world's factory" for low-tech manufacturing into a "smart device" global power. However, the US sanction on ZTE, which relied on American suppliers such as Qualcomm and Lumentum for about one-third of its components, and its abrupt operation suspension within weeks, highlighted the tech power gap still existing between China and the United States.

The Trump administration's ZTE move was not a knee-jerk reaction to China's rising tech power; instead, it was part of a strategic response to maintain America's innovation supremacy. Amid tech export restrictions, China has stepped up efforts to bolster its own chip industry through various means, including the acquisition of foreign technology companies, but the effort encountered pushback by the US **Committee on Foreign Investment in the United States (CFIUS),** the agency that screens foreign direct investment for national security risks.

With an eye to further constraining foreign incursions in the sensitive tech sector, since 2018 (effective in early 2020) the Trump administration has developed new CFIUS rules to scrutinize Chinese and other foreign investments more carefully, especially in high-tech fields. The new rule emphasizes national security review of transactions involving "TID" technologies—critical technology, critical infrastructure, and sensitive personal data (see Figure 9.1).

As this chapter shows, while the rules would apply to any foreign investment, the effort is likely aimed at preventing China from gaining access to sensitive American technology—after years of Chinese companies acquiring startups and cutting-edge emerging companies in Silicon Valley (and innovation hubs worldwide). At the same time, these US actions have firmed up China's resolve to cut reliance on US tech smarts and to grow its own core technologies. The battle for digital tech supremacy is on.

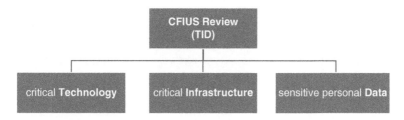

Figure 9.1   The TID focus of CFIUS

TID – Not All National Security Is the Same

The concept of "national security" is a slippery one, subject to change with the vagaries of public reaction. In the beginning, "national security" was generally seen as synonymous with military security, involving military suppliers, technology, and facilities. It has since ballooned to encompass much more. The United States is a bellwether in this expansion of "national security" into surprising corners of the economy.

In the United States, **CFIUS** which we discuss in detail in the following text, reviews and decides on most investments. Occasionally, presidential actions may also be informed by the recommendation of CFIUS. Since its creation, presidential action has blocked seven transactions based on CFIUS recommendations, with the tempo noticeably increasing, from President Barack Obama onward, in the last five years—along with the global tech revolution, more digital economy transactions were reviewed and challenged; coincidentally, all seven of them involved Chinese buyers (or a China factor, in the Broadcom/Qualcomm case). The short list of presidential

Table 9.1   [Seven] Six transactions blocked by US Presidential action

Year	US President	Investor	Target
1990	George H.W. Bush	China National Aero-Technology Import and Export Corporation (CATIC)	MAMCO Manufacturing
2012	Barack Obama	Ralls Corporation (a China-controlled company)	Oregon wind farm project
2016	Barack Obama	Fujian Grand Chip Investment Fund	Aixtron, a German-based semi-conductor firm with US assets
2017	Donald Trump	Canyon Bridge Capital Partners, a Chinese investment firm	Lattice Semiconductor Corp. of Portland, OR
2018	Donald Trump	Singapore-based Broadcom	Qualcomm, a US semiconductor chip maker
2020	Donald Trump	Beijing Shiji Information Technology Co., Ltd	StayNTouch, a US software supplier to the hospitality industry
2020	Donald Trump	Bytedance (Beijing)	Musical.ly (which created TikTok in its current form)

actions provides a glimpse into that evolution of "national security" reviews.

The Committee on Foreign Investment in the United States (CFIUS) dated back to a 1975 executive order from President Gerald Ford. Ironically, CFIUS was established to "dissuade Congress from enacting new restrictions" on foreign direct investment. Ford's predecessor in office, Richard Nixon, under the pressure of OPEC-induced concerns over the weakness of the US dollar, had been forced to suspend the convertibility of dollars into gold. And the increasing investment by the oil rich Gulf States, in particular, was seen as politically—not commercially—motivated. The creation of CFIUS appeared to have appeased these concerns by instituting a review mechanism consisting (today, after expansion) of nine cabinet members and other officials in the executive branch of the US federal government (see Figure 9.2).

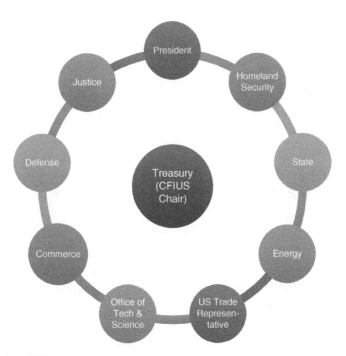

Figure 9.2   CFIUS member agencies

CFIUS takes the form of an interagency committee headed by the Secretary of the Treasury and includes the Secretaries of Commerce, Defense, Energy, Homeland Security, and State, as well as the Attorney General, the US Trade Representative, and the Director of the Office of Science and Technology Policy. In addition, the Secretary of Labor and the Director of National Intelligence are non-voting members, and five White House offices hold observer status. Treasury staff handle the administration, and all communications with parties to a CFIUS review are conducted through them. Decisions are reached by consensus. Reviews were conducted without public disclosure, and for decades, CFIUS operated in relative obscurity.

CFIUS has evolved in recent decades, and the ever-expanding rationale for "national security" review doubtlessly contributed to the increased number of submissions and reviews. Still, until the 2016 Presidential elections, the number of transactions reviewed remained limited and the number of outright rejections (not counting deals abandoned due to CFIUS scrutiny) remained a handful. The numbers, however, spiked in 2017 as the Trump administration dramatically expanded the use of CFIUS reviews.

Under the Trump administration, CFIUS appears to have "stepped up" restrictions on foreign investments for "national security" reasons. As America's technological hegemony is under threat from China, the remit of CFIUS is being significantly expanded to cover all transactions in the digital economy. "The digital, data-driven economy has created national security vulnerabilities never before seen", Heath Tarbert, assistant secretary of the Treasury for international markets and investment policy, said in a prepared testimony to the Senate Banking Committee. "Today, the acquisition of a Silicon Valley startup may raise just as serious concerns from a national security perspective as the acquisition of a defense or aerospace company".

The Qualcomm Fighter Jet Scramble

On March 12, 2018, President Trump blocked Singapore-based Broadcom Limited's US$117 billion bid for Qualcomm Incorporated. Broadcom, a US tech company based in Singapore for tax considerations, was viewed as an aggressive cost cutter. It was feared that the acquisition would lead to severe cuts in the R&D investments at Qualcomm, and hence, America's loss of the battle to dominate fifth-generation (5G) wireless technology where China's Huawei is in the lead.

The Treasury chair of CFIUS even sent an unprecedented letter to the parties in the CFIUS proceeding in March 2018 to this effect. Complicating matters, Broadcom was in the process of redomiciling as a US entity, which would have arguably deprived CFIUS of jurisdiction. CFIUS reacted by requiring Broadcom to provide advance notice before so proceeding—presumably to allow CFIUS the power to block the deal. It all added up to a sweeping land grab by CFIUS where many viewed the transaction as not involving national security, but rather a blatant effort to favor the home team in the 5G race.

The Qualcomm fighter jet scramble made clear that the Trump administration's concern about Huawei's leading role in 5G wireless technology. The no-holds-barred nature of the reaction served to lay bare that the United States is strategically disadvantaged in the 5G battle by the lack of a US champion. If CFIUS is one of the arrows in the quiver, interestingly, the concept of state investment funds—better known in the China context—is another. Attorney General Bill Barr has suggested a US national fund to acquire a controlling stake of the European firms, and Congress is separately contemplating a 5G tech-specific fund to develop alternative products to Huawei's 5G network.

As the cases in Table 9.1 show, even before Congress acted in the summer of 2018 to codify the new rules of the game, a distinct trend in US Presidential actions emerged. The force of the reaction to foreign direct investment in the United States became evident when the White House and both parties, in one of the most politically divisive moments of US history, circled the wagons and fortified the role of CFIUS and its scrutiny of foreigners buying in the United States. Playing catch up with the transformative Trump moves, Congress enacted the Foreign Investment Risk Review Modernization Act (**FIRRMA**) in 2018, and the final rules became effective in February 2020.

FIRRMA largely codified the practices that had become a fait accompli with the Qualcomm (see the "The **Qualcomm Fighter**

Jet Scramble" box) and Moneygram (see the case study in a later section) blockages. While the entire CFIUS process was still opaque, FIRRMA articulated for the first time CFIUS practices that, under the Trump administration, CFIUS had already been pursuing in the digital economy. CFIUS, in its more robust oversight role, was targeting Chinese-related investment in the United States, though not exclusively. The new legislation and regulations adopted since 2018 have made those security review-based investment barriers formal and permanent. As the following sections illustrate, the investment restrictions have become part of the "new normal" in the US–China tech war.

Critical Technology and Semiconductor Chips

As mentioned earlier, the new FIRRMA legislation in the United States mandates that CFIUS view national security through a wider lens: in addition to the more traditional indicia, such as protection of defense facilities and infrastructure, government contracts, and so on, the innovation capability to develop emerging "TID" technologies is the new focus:

- **critical technology:** including the sorts of military- and defense-related items with which CFIUS has traditionally been associated, as well as certain "emerging and foundational technologies" used in industries such as computer storage, semiconductors, and telecommunications equipment.
- **critical infrastructure:** identified by reference to a list of 28 subsectors, including, among others, telecommunications networks, electric power generation, transmission, distribution and storage facilities, certain oil and gas systems, financial market utilities and exchanges, and airports and ports.
- **sensitive personal data:** include any business that maintains or collects genetic information or other "identifiable data", such as financial, health-related, biometric, or insurance data for more than one million individuals.

Leading-edge technology has always been a major focus of the US CFIUS for "national security" analysis. Back in 1987, the Japanese electronics firm Fujitsu sought to buy 80% of Fairchild Semiconductor, a US semiconductor producer, from Schlumberger, a Franco-American NYSE-listed company. Fairchild to Intel and many other US semiconductor companies was like the more recent Netscape to today's Google and Facebook, and a *Los Angeles Times* news story then described it as "a pioneer of America's high-technology industries" and the "mother company of Silicon Valley".

The deal was viewed by many in the Reagan administration and industry officials as a test case in Japanese efforts to invest in the United States, particularly in strategically important high technology. Fairchild was a military supplier to the US armed forces, and then Defense Secretary Caspar W. Weinberger, Commerce Secretary Malcolm Baldrige, and the Central Intelligence Agency (CIA) collectively asked the White House to block the Japanese electronics company's bid. The *Los Angeles Times* suggested that the potential transaction with "foreign rival Fujitsu of Japan" is like "selling Mount Vernon to the redcoats". Under intense pressure, the parties called off the deal.

History rhymes. In 2016, capital from China made a US$2.49 billion acquisition offer to the same Fairchild Semiconductor company (see Figure 9.3). The buyer consortium consisted of China Resources Microelectronics Ltd., a state-owned

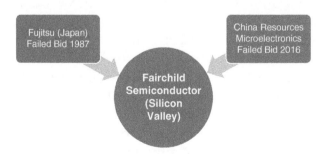

Figure 9.3   Japan and China's bids for Fairchild

company, and Hua Capital Management Co., a China-based investment fund (whose LPs probably also included a China state investor). The Chinese well understood the deal might be tied up in a lengthy national security review; hence, the consortium promised to pay the company US$108 million if the deal failed to clear the bar with CFIUS. Nevertheless, the Fairchild board decided the proposed deal wasn't worth the regulatory headaches (again), and it rejected the offer, citing concerns over the uncertainty of obtaining CFIUS approval.

The same year, 2016, also saw the launch of a deal that later fell afoul of CFIUS. President Trump blocked the Chinese government-backed PE firm Canyon Bridge from buying US chip maker Lattice Semiconductor Corporation for US$1.3 billion. As is discussed in detail in Chapter 1, Oregon-based Lattice is a publicly traded semiconductor company that manufactures programmable logic devices that can be used by customers for specific uses such as vehicles, computers, and mobile phones.

One of the main issues arising in the Lattice/Canyon Bridge Deal was the strong connection with the Chinese government. Lattice started seeking potential acquirers in February 2016, and on April 8, 2016, China Reform Fund Management Co ("**CRFM**") approached Lattice for a potential takeover bid. CRFM owned businesses with military applications and also invested in other sensitive industries in line with the Chinese government's military and economic agendas. As such, CRFM's bid would naturally be subject to a US CFIUS review.

Probably realizing this problematic tie to sovereign capital, CRFM designed a complex transaction structure by establishing a US-based private equity fund, Canyon Bridge Capital Partners, Inc. (see Figure 9.4). China Venture Capital Fund ("**CVCF**"), a subsidiary of CRFM, was the sole limited partner of Canyon Bridge, and two US nationals (with CRFM affiliation) were appointed as the general partner (GP) to manage the fund. This structure aimed to show that the potential purchaser, Canyon Bridge, is a US-based entity and is operated by a US-based general partner and CRFM, as the limited

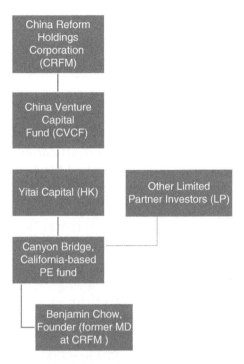

Figure 9.4   The chain of ownership and control of Canyon Bridge

partner, is only acting as a passive investor with limited operational rights.

Still, President Trump in September 2017 issued an executive order blocking Canyon Bridge (CRFM) from buying the chipmaker Lattice Semiconductor. According to his order, "the national security risk posed by the transaction relates to, among other things, the potential transfer of intellectual property to the foreign acquirer, the Chinese government's role in supporting this transaction, the importance of semiconductor supply chain integrity to the US government, and the use of Lattice products by the US government".

The Lattice Semiconductor case constituted the second Chinese connected acquisition in the semiconductor industry to be blocked within a year, sending a signal that the United States will specially oppose China-related investments in the high-tech sector. (Months earlier, in December 2016,

President Obama had blocked a Chinese acquisition of the US business of German semiconductor company Aixtron SE. The US assessment identified that Aixtron's US subsidiary not only to include its corporate presence, that is, its research, technology, and sales facility in California, but also the company's US patents, granted and pending. The United States took the far-reaching position that the existence of US intellectual property may be considered an independent ground for CFIUS jurisdiction, which is critical for future transactions in the digital economy.)

Furthermore, the case also reflected how the United States now employs the review process: no matter how the purchaser designs the deal structure, CFIUS will eventually trace back to the ultimate controller and its affiliates (in this case the CVCF capital had even passed through a HK-based vehicle before reaching the US-based Canyon Bridge; see Figure 9.4) to determine whether the target company will be controlled by a foreign buyer after completion.

More recently, in February 2018, CFIUS in effect blocked the US$580 million acquisition of semiconductor testing company Xcerra by Hubei Xinyan, a Chinese state-backed semiconductor investment fund. (Technically, Xcerra withdrew in the face of signals that CFIUS would never approve the deal.) Xcerra does not manufacture chips itself. Instead, it provides testing equipment used in the production of semiconductors. The deal was terminated due to CFIUS's concern that Xcerra's equipment was used by chip manufacturers that were part of the supply chain to the US government.

From Hard Tech to Invisible Data

Now in the age of 5G and iABCD, with the new Foreign Investment Risk Review Modernization Act ("**FIRRMA**") enacted in 2018 and fully in effect as of February 13, 2020, the concept of "critical technology" has been substantially expanded (along with significant reforms to CFIUS, as discussed in previous section). CFIUS traditionally associated "critical technology" with military- and defense-related items, but the new regulations also highlight "emerging and foundational technologies" that could be discovered and applied in industries as wide-ranging as computer storage, semiconductor electronics, and telecommunications equipment.

In that context, data (the "D" of the "TID" categories mentioned at the beginning of the chapter)—whether personal or industrial—has emerged as a mainstream concern of CFIUS. The background: AI must be "trained" or "taught" for its cognitive abilities. From facial recognition to autonomous cars to machine translation, most AI applications must digest vast amounts of data and find hidden patterns between inputs and outcomes (so-called "machine learning") before a machine algorithm could "learn" how to master human skills. (See Chapter 7 for related discussions on China's data industry.)

Therefore, to a significant degree, the AI innovation race is a competition for data; in addition, data collection and analytics also raise serious personal privacy issues. Lately, data technology related startups have become the new focus of CFIUS. This section analyzes related high-profile cases, including three Chinese companies being subjected to the previously rare remedy of retroactive divestment of technology startups Grindr, PatientsLikeMe, and StayNTouch in 2019 and 2020, allegedly due to concerns regarding access to sensitive personal data.

First, in March 2019, CFIUS directed the Chinese gaming company Beijing Kunlun Tech to sell its stakes in the popular gay dating app Grindr. Kunlun took over Grindr through two separate transactions between 2016 (61.53%) and 2018 (100% buyout). In March 2019, spurred by data privacy concerns, CFIUS intervened. (Kunlun completed the earlier deals without submitting the acquisition for CFIUS review because the filing requirement was voluntary and probably in light of the fact that a gay dating app was not thought by the venture capital world to be a national security risk.)

In March 2020, Kunlun announced the sale of Grindr to San Vicente Acquisition for roughly US$600 million, ahead of the June 2020 divestment deadline. (The transaction was subsequently closed in June 2020.) The sale reflected a substantial increase over the US$151 million valuation at which Kunlun had acquired its majority stake in 2016, but likely reflects a discount from its true value due to the forced sale.

Previously, in 2017, Ant Financial, Alibaba's financial arm, announced a US$1.2 billion acquisition of money transfer company MoneyGram. Ant Financial had a strategic view that MoneyGram could provide a platform for Chinese Alipay customers to access goods and services when traveling abroad. However, Ant Financial was not able to mitigate CFIUS concerns over the safety of data that could be used to identify US citizens, and the deal collapsed in January 2018.

In the Grindr case, CFIUS has not disclosed its specific concerns, but "data privacy" apparently is the main concern. Grindr, as described earlier in this chapter, gathers sensitive personal data. According to some analyses, the Grindr users (including US military and government contractors) could be blackmailed by their social data.

Second, more than a year after the investment, CFIUS ordered the Chinese tech company iCarbonX to sell its majority stake in PatientsLikeMe. The reason apparently related to the collection and archiving of US patient data by PatientsLikeMe. A US health tech startup company, PatientsLikeMe is an online service that links individuals by providing a social network on which patients can connect to people with similar health conditions. In 2017, iCarbonX, the Shenzhen, China-based digital healthcare unicorn, invested more than US$100 million in PatientsLikeMe, becoming the majority stakeholder. In mid-2019, iCarbonX fully divested when the startup was acquired by the US company UnitedHealth Group Inc.

Third, StayNTouch/Beijing Shiji Group became the third Chinese purchase involving personal data retroactively undone by US government's order (this time by Presidential

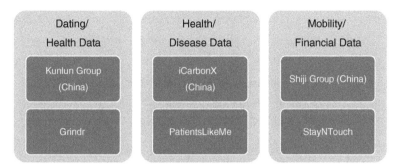

Figure 9.5   PII transactions undone by National Security Review

order, making it one of the only six transactions blocked by presidential action based on CFIUS recommendations). Beijing Shiji Group, a Chinese software supplier to hotels, restaurants, and retail, in 2018, acquired a StayNTouch, a US-based company providing cloud-based software to the hospitality sector, including MGM Resorts International, Yotel, and Miami Beach's Fontainebleau, famously featured in *Goldfinger*, an early James Bond film. The powerful finger's touch is in play here—the software enables hotel guests to check in and check out using their smartphones.

In March 2020, years after the deal closed, Beijing Shiji Group was directed to divest StayNTouch and its US assets within 120 days of the order. As with Grindr, the StayNTouch divestment order does not specify what element raised national security concerns. The order provides a clue, however. Despite the 120 day window to divest, the Chinese company and its Hong Kong subsidiary are ordered to immediately refrain from accessing guest data. This element supports the thesis that Chinese access to personally identifiable information (**PII**) of US users, no matter the industry, is now a national security risk in the view of the Trump administration (see Figure 9.5).

The latest additions to the list are ByteDance and TikTok (see related discussions in Chapters 1 and 4.) In 2019 CFIUS reportedly commenced an informal investigation into the long

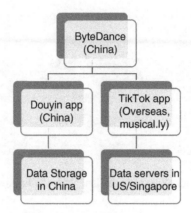

Figure 9.6   ByteDance ring-fenced data for CFIUS

concluded Chinese acquisition of the wildly popular app musical.ly. In 2017, Chinese tech powerhouse ByteDance paid US$1 billion for the startup musical.ly—the originator of the now viral social video app TikTok. CFIUS' approval was not sought at the time, leaving the deal vulnerable to a subsequent divestment order from the agency. The acquisition succeeded spectacularly, which led to serious personal data issues in the United States—especially relating to teenagers and US government employees.

ByteDance has a plan to satisfy the concerns without divesting. In reaction to the informal CFIUS probe, ByteDance has reportedly ring-fenced data from US users, keeping it on servers in the United States and Singapore, not in China (see Figure 9.6). In another countermove, TikTok has hired its first "chief information security officer". Roland Cloutier, who began his career with the US Air Force as a combat security specialist and served in senior security capacities at US public companies, burnishes the app's US security credentials; in 2016, he literally wrote the book on his new role: *Becoming a Global Chief Security Executive Officer*.

Also in 2020, TikTok opened a "transparency center" in Los Angeles, where it promises to share information about its content-moderation, privacy, and security controls, and said it would stop using moderators in China to handle content from users outside the country. Moreover, in May 2020, Disney's Head of Streaming, Kevin Mayer, was recruited to become TikTok's new CEO. According to media reports in July 2020, TikTok also agreed to buy more than US$800 million in cloud services from Google.

TikTok evidently hoped the measures taken to separate its Chinese and US operations and safeguard personal data of US users could save the lip-synching video app from a CFIUS divestment order, years after the deal closed. Still, in August 2020, US President Trump issued an executive order whereby ByteDance must divest its US operations of TikTok. Microsoft and Oracle and other private investors immediately started a bid for TikTok's businesses in the US, Canada, Australia and New Zealand, and Meyer soon resigned from his new job. To emphasize the significance of the personal data involved, US Treasury Secretary Steven Mnuchin stated that the order also requires ByteDance to divest "any data obtained or derived from TikTok or Musical.ly users in the United States."

China and the United States—Mutually Self-sufficient?

The years of 2019–2020 may be remembered as an important inflection point for the global digital economy, from collaborative co-existence to head-on tension, as illustrated by the selective blocked cross-border transactions between the United States and China. With afterthought, the disrupted overseas investments, extended tech export blacklist, plus the ongoing trade and tech wars with the United States, may have provided China with fresh impetus to accelerate its efforts to gain self-sufficiency.

China still has many weak areas to catch up on, for example, its digital economy is dependent on US software, from Microsoft's Windows operating system to Google's mobile apps for Android smartphones. However, the largest gap is in the design and manufacture of semiconductors and high-end chips, which is the foundation for the country's vast manufacturing supply chain, as illustrated by the ZTE case. Consequently, China is mobilizing a national effort to seize the global lead in chip technology in the same way as the AI national strategy (see Chapter 1 discussions), with tens of billions of dollars in new government investments.

In October 2019, China set up a new national semiconductor-focused fund, the China National Integrated Circuit Industry Investment Fund (the second in less than five years), with 204 billion yuan (US\$28.9 billion)—its predecessor was capitalized with US\$20 billion in 2014. The fund's registered capital comes mainly from state organizations, according to company registration information, which included China's Ministry of Finance (22.5 billion yuan) and the policy bank China Development Bank (22 billion yuan) as well as state-owned enterprises such as China Tobacco Co.

The new fund plays to the tune of domestic tech development to reduce dependence on strategic imports and accelerate the digital transformation of the economy. But in semiconductors, China still faces a long path to global leadership. The 2014 fund poured billions of dollars into dozens of projects, but its electronic products still lag behind industry leaders. Taiwan and South Korea both possess advanced chip production capacity, although expertise in the manufacture of capital goods for the fabrication of chips resides in Western Europe, Japan, and North America.

The new fund's goal may become even broader—to cultivate China's complete semiconductor supply chain, from chip design to manufacturing, from processors to storage chips. Naturally, it has attracted fresh concern from the US officials, who complained the new fund—just like its 2014 predecessor—amounted to a form of state capitalism that gave its companies an unfair advantage against its US competitors. Also, the US government has an interest in the security of the supply chain in this sector because many semiconductors are ultimately used by the US government and military.

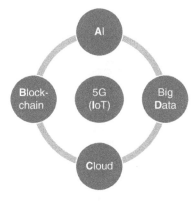

Figure 9.7   5G-powered iABCD

The United States' response? Its own sovereign funds on 5G network technology developments, an area that the United States and its Western allies rely on China-sourced products—especially from Huawei. Huawei is paradigmatic of the challenge China poses the United States. It is a technological leader in "5G" telecommunications, the next-generation network technology that will enable new industries like IOT, AI, blockchain, cloud and Big Data (see Figure 9.7). The desperation in the United States was underscored by the highly unorthodox Trump maneuver to deploy CFIUS to preserve Qualcomm's R&D budget, the likely best route to developing a US champion in 5G. Meanwhile, Huawei has been expanding the contest beyond 5G to the internet itself.

The Trump administration's diplomatic campaign to persuade Western countries (as well as US domestic carrier operators) to shun Huawei's 5G network has had only mixed success, due to the need to develop viable alternatives to Chinese products. Australia has signed on, and the UK government announced in mid-2020 that it would bar the purchase of new Huawei equipment for 5G networks after December, and that existing gear already installed would need to be removed from the networks by 2027. However, other countries have Huawei as their telecoms supplier.

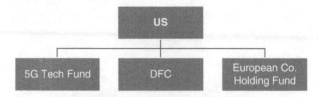

Figure 9.8   New sovereign funds of the United States for 5G

The inability to gain traction in the push to avoid Huawei products is hardly surprising, given the actual market conditions. Competitors such as Europe's Ericsson and Nokia often do not offer the type of fully integrated solutions that Huawei does, and they come at a higher upfront cost. Countries, particularly poorer emerging markets, are reluctant to slow down 5G deployment because they see rapid deployment of 5G as important to their own economic growth. In a world of limited and high-cost alternatives, governments will be tempted to turn to Huawei despite security risks.

As a result, the United States has championed alternatives to Chinese tech (see Figure 9.8). One took the form of a program at the newly created U.S. **International Development Finance Corporation (DFC)**. The DFC (the successor to OPIC, the Overseas Private Investment Corporation), with US$60 billion in funding, announced in December 2019 a program to provide finance to emerging markets in the development of mobile networks, *provided that* they do not employ Huawei equipment. Although it is not clear what alternative equipment suppliers would be tapped, Bloomberg reported at the time that shares of both Nokia and Ericsson rose with the announcement.

In January 2020, the concept of another US sovereign fund to counter China's moves was floated. During the very same week that China and the United States signed their phase I trade war "truce", a bipartisan group of US senators introduced legislation that would provide over US$1 billion to fund the development of Western alternatives to Huawei. The bill proposes that the Federal Communications Commission (**FCC**) direct at least US$750 million or up to 5% of annual auction proceeds from newly auctioned spectrum licenses to create an open-architecture model (Open Radio Access Network, or "O-RAN") research and development fund. Grants from the fund would be overseen by the National Telecommunications and Information Administration (**NTIA**) with input from other agencies. The potential recipients are not limited to US companies.

Finally, speaking in February 2020, Attorney General Bill Barr offered up the third US fund in an even more blatantly interventionist approach. Despite the efforts to launch Qualcomm on the path to 5G dominance, most observers would conclude that the United States does not have a homegrown competitor to Huawei. The only viable rivals are two European telecoms firms. Barr is quoted: "Some propose that these concerns could be met by the United States aligning itself with Nokia and/or Ericsson through American ownership of a controlling stake, either directly or through a consortium of private American and allied companies". Maybe this is where the senators would invest the US$750 million fund proposed the preceding month, or the US$60 billion DFC war chest, or capital from other US government sources.

Europe, Japan, and More—the Divided Tech World

As China and the United States compete for world dominance in tech innovation, other innovation hubs such as Europe and Japan are taking mixed measures. On one hand, several advanced EU economies comparable to the United States have recently made regulatory changes, similarly increasing their focus on foreign investments (especially China) in high tech and digital economy sectors. Particularly relevant countries include Germany, the United Kingdom, and France. On the other hand, they are pursuing their own innovation strategy as an independent pillar in the global digital economy. The same it also true for Japan.

In **Germany**, the 2016 takeover of Kuka, a Bavarian robotics firm, by a Chinese SOE was described by some German officials as a "wake-up call" that underlined the need to shield strategic parts of the economy. In November 2019, Germany announced its plan to tighten scrutiny for foreign investments by formally amending foreign investment rules for critical sectors. The plan would expand "critical industries" to sectors including artificial intelligence, robotics, semiconductors, biotechnology, and quantum technology, and requiring disclosure for any purchases over 10% of companies in those areas. Meanwhile, for other sectors that had previously been identified and regulated as "critical infrastructure", including energy, telecommunications, defense, and water, the percentage threshold that triggers a government review was lowered from 25% to 10%.

In the **United Kingdom**, new regulations that took effect in June 2018 lowered merger control thresholds at which the government would be able to intervene, affecting advanced technologies, including computer hardware and quantum technologies, the military, and dual-use sectors. The wide-ranging changes to the national security landscape, which were originally expected to have been published in 2019, have not yet been published. It is anticipated that changes will not be implemented until sometime in 2020.

France also took similar regulatory measures in November 2018. The list of sensitive activities was further expanded, and the government review now covers activities relating to "the interception/detection of correspondence/conversations, capture of computer data, information systems security, space operations and electronic systems use in public security missions". The scope of review is further widened to cover research and development (R&D) activities in "cybersecurity, artificial intelligence, robotics, additive manufacturing, semiconductors, certain dual-use goods and technologies, and sensitive data storage".

In 2019, **Japan**'s government expanded the scope of the industries for which inward direct investments require advance notice filings and government review. Artificial intelligence (AI) is a major focus of the revised rules. Industries pertaining to hardware (e.g. integrated circuits) and embedded software, which are essential for the application of AI technologies, have been added to the review list. Such a move seems to suggest that Japan intends generally to require all industries pertaining to AI developments be subject to the prior notification requirement. Overall, it is hardly surprising that Japan's new rules mirror the US FIRRMA in many aspects, both in contents and consequences.

In April 2020 amid the COVID-19 crisis, the EU competition chief, Margrethe Vestager, publicly called on EU member states to accelerate the adoption of regulations to fend off Chinese SOEs from buying European companies whose share prices have fallen due to the pandemic. Not only that, but she encouraged governments to take stakes in companies if needed to prevent them falling into the clutches of a Chinese SOE. Italy is not waiting on anybody, having already adopted increased measures to prevent foreign takeovers. Germany and Spain similarly announced anti-takeover initiatives. More countries like Ireland started to examine stricter foreign investment controls to protect strategic sectors (such as critical technologies) that may have become especially vulnerable as a result of the COVID-19 disruption.

Meanwhile, Europe is taking steps to develop its own innovation power. The European Union in February 2020 unveiled a plan to restore what officials called "technological sovereignty", with more public spending for the European tech sector. With the global economy becoming ever more reliant on digital technology, European leaders are concerned that the European economy is overly dependent on technology developed and controlled elsewhere. As Ursula von der Leyen, the president of the European Commission, the executive branch of the EU, said at a news conference, "We want to find European solutions in the digital age".

Indeed, there are plenty of areas that could use European solutions. For Europeans, Apple of the United States and Samsung from South Korea are the most popular phones. Similarly, US companies dominate digital platforms in Europe: Facebook operates the most widely used social networks; Google rules online search and advertising; and Amazon reigns over e-commerce. Cloud infrastructure from Amazon and Microsoft is indispensable to European companies. Meanwhile, China's Huawei produces the physical equipment on which Europe's digital economy runs. Driving home the extent of foreign digital dominance, EU officials had to call Los Gatos, California, to ask US tech giant Netflix to lower its video streaming quality to prevent a European system crash during the coronavirus-induced surge in internet traffic.

Hence, a government investment fund—like its counterpart of China and the United States—is now in talks. According to media reports in August 2019, EU staff has drafted a plan to launch a €100 billion (US$110 billion) sovereign wealth fund, to be called the "European Future Fund". The main goal of this proposed fund will be to invest in future "European tech champions", which could potentially compete in the same league as China's BAT (Baidu, Alibaba and Tencent) or the US GAFA (Google, Apple, Facebook, and Amazon). Due to the complex EU politics, it is not clear that the fund will ever be realized, but the determination to compete with American and Chinese tech dominance is clear.

Also joining in the 5G sparring match between China and the United States, Japan, in 2020, is reportedly stepping into the ring by appropriating 220 billion yen (approximately US$2 billion) for corporate research and development of "6G" technology (the next generation telecommunication tech after 5G—not a typo here). Japan's new fund will be included as part of an economic stimulus package, and the state-run New Energy and Industrial Technology Development Organization (NEDO) will house it. In February 2020, the Japanese government announced that it is working on a detailed blueprint focusing on 6G wireless communications network with a dedicated panel moving discussions on technological developments, potential utilization methods, and policies forward for operations in 2030.

Having lagged behind China, Korea, and the United States in the introduction of 5G, Japan is funding its efforts aggressively in order to get back in the race for 6G and aims to play a key role. "The smooth introduction of standards for next-generation wireless communications networks is indispensable to boosting Japan's international competitiveness", Japan's communications minister, Sanae Takaichi, said at the news conference accompanying the announcement of the 2030 goal.

In summary, more and more countries are joining the 5G (and 6G), AI, and semiconductor race for international competitiveness. Ironically, the internet of things (IoT) has not only spurred the governments of the world into the tech-driven developments, but also spawned new state-investment-funds to serve as aggressive policy tools in carving up the digital future by funding its fracturing.

No Winner in Balkanization of Cyberspace

In 2020, the tech war shows no sign of slowing down. In May 2020, the US Commerce Department further tightened its regulatory vise on Huawei. In the year before, the US government blocked the export to Huawei of American-made chips; now it's declaring that even chips made with US equipment—anywhere in the world—can't legally be sold to Huawei. The rule, still subject to revision, means that if a foreign semiconductor manufacturer (so-called foundries) fills a Huawei order using US technology, it will be a violation of the extraterritorial order. The impact is on the whole semiconductor supply chain because American software and equipment is behind all foundries that fabricate advanced chips.

On the offensive side, in June 2020, US lawmakers proposed an estimated US$25 billion in funding and tax credits to strengthen domestic semiconductor production and counter rising technological competition from China. Around the same time, the United States convinced Taiwan-based TSMC, the largest semiconductor maker in the world, to announce that it would build a major foundry in Arizona. TSMC gets about a sixth of its revenue from Huawei, but it may need to cut this important client due to US law compliance. For the

United States, the TSMC–Arizona joint venture not only pulls high-end chip manufacturing out of China's reach, but also establishes a leading semiconductor foundry on the US land.

In the semiconductor industry's globally integrated network, policies aimed at decoupling will hurt everyone. Banning US and global companies, such as Intel, AMD, Qualcomm, Broadcom, and Texas Instruments from doing business with Chinese players will indeed delay the development of the semiconductor industry in China, but it also hurts American and other companies' competitiveness. The semiconductor equipment leaders are American, Japanese, and Dutch companies, and China's "domestic" microchip production without their equipment is almost impossible.

Meanwhile, these companies, as well as the foundries of TSMC and Samsung, also need their revenue from China to be able to invest in R&D and keep innovating. As mentioned in an earlier section of this chapter, the "other" countries may face an especially difficult choice when they are caught up in the China–US crossfire. More companies in those countries are seen to increasingly locate more research and development outside the United States to ensure that they have continuous access to China, a vast and growing consumer tech market (such as smartphones and wearable devices) and the center of the global electronics supply chain. Nevertheless, there is no guarantee that such best practice would last long.

All in all, a US–China tech decoupling is real and accelerating. Hence, the digital economy is in a vital conflict and crisis: the global tech world, and together with at least part of the world economy, is now fractured into two—and potentially more, considering Europe, Japan, and other regions—spheres of influence, whereas tech entrepreneurs are driving the prospect of a technological singularity, hyper-connected society, and internet of everything. What's hopeful is that the economic relationship and innovation interdependence among countries may create a new equilibrium and a shared digital future. That is why the Digital Silk Road is such an important concept to start the global digital economy dialogue for a shared future.

CHAPTER 10

The Silk Road in Cyberspace

- Global Connectivity, Digital Divide
- DSR Greens the Belt and Road Initiative
- Farmer Retailers, Taobao Villages, and the Alibaba University
- Link SMEs with Mega Platforms
- AI Anxiety and DSR Co-Development
- Israel and Everyone: Caught in the Middle
- Innovative, Invigorated, Interconnected, and Inclusive

Global Connectivity, Digital Divide

The internet and social media now seem to connect everyone to everyone else, and to make information available to all. But even in 2020, "everyone" is a bit of an exaggeration. More than 3 billion people, or roughly half the world's population, still have no access. Which brings me to my question: How long will it take to bring the next billion users into the digital fold?

The answer may sound pessimistic. But without a very determined effort, the process of networking the globe may take much longer than people expect—a delay the world can ill afford. Why? All you need is to look at is the Chinese experience to appreciate the challenges to come.

This may sound counterintuitive in the final chapter of this book. This book (as well as its predecessor, the 2016 *China's Mobile Economy*) celebrates the world's largest internet user population: more than 900 million Chinese are now online, almost as many as the combined population of the United States and the European Union. The 300 million-strong digitally connected Chinese middle class has redefined the consumer industry in China—and, for that matter, in the world. However, this consumption boom in China built on digital technology also illustrates the challenges involved in the sharing of cyberspace-driven prosperity.

Thanks to China's encouragement of urbanization and massive infrastructure investments, most of the 900 million users (99% of whom connect by means of mobile devices) have been acquired at a remarkable pace over the last decade. However, the country still has hundreds of millions non-users to convert. Indeed, China's internet penetration rate just reached 64.5% by 2019, which puts the country only modestly ahead of the world average. What's more, the growth in connections has tapered off in recent years, in large part, because penetration in rural areas is still less than half that of the cities (see Figure 10.1).

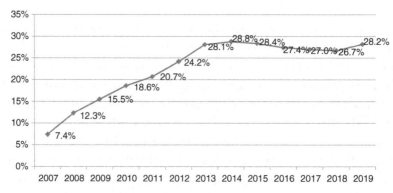

Figure 10.1   Internet users in rural areas as % of total Internet Population in China

Hundreds of millions of Chinese thus lack the benefits of applications ranging from social networking to online shopping to access to a cornucopia of entertainment. And the poorer people are, the more a smartphone outperforms all the other options they can afford to turn otherwise empty time into something enjoyable. Households in the rural areas tend to have a single television shared by large families. For the teenagers, the ability to consume media of their own choice is a sea change from having to watch whatever grandpa has chosen.

But future internet penetration in rural China will require substantial investments in the network infrastructure in remote areas—investments that may be delayed without financial and organizational support from the central government. That aid meshes well with the government's programs such as "Development of China's Western Region" that focuses on infrastructure investments in remote areas. Compared to China, the infrastructure needs and financing gap are even more significant in the emerging markets, where the young generation's quest for communication, entertainment, and self-expression are the same.

People want to stay in touch with each other, to be entertained, and to express themselves, which is true in the rich world and in China as well as everywhere else. Throughout the world, the young generation's native language in the cyberspace is video. They watch videos—which they are also making, in great abundance. On the one hand, with few voiceovers, this content easily travels globally. On the other hand, video is easier to post to one's peers than writing (language) is.

In short, there are still striking cross-country and cross-regional differences in internet penetration rates, infrastructure construction, technological innovation, and digital literacy all over the world. With the global digital economy developing at a breathtaking pace, multinational cooperation to accelerate coverage in places that have yet to be swept up in the digital information revolution should be an equal priority. Providing access to entertainment contents, virtual communities for a richer social life, and the ability to express and be heard (and seen) to hundreds of millions will mark a profound improvement in next generation's aggregate quality of life. That's where the "Digital Silk Road" (**DSR**) fits in—and starts with.

At the December 2017 World Internet Conference at Wuzhen, China, China's DSR cooperation initiative was officially launched and seven countries including Saudi Arabia, Egypt, Turkey, Thailand, Laos, Serbia, and the United Arab Emirates signed up as the first group. By the second Belt and Road Forum (BRF) in April 2019, the list of countries that had signed DSR-specific MOUs (memorandums of understanding) with China extended to 16 from almost all major continents (see Table 10.1). (As mentioned at the beginning of this book, the Digital Silk Road concept was officially coined by China's President Xi at the first BRF in 2017.) In addition, the governments of Austria, Argentina, Chile, Brazil,

Table 10.1   Countries signing DSR-specific MOUs with China

	Region	Country
1	Africa	Egypt
2	Asia	Turkey
3	Asia	Bangladesh
4	Asia	Laos
5	Asia	South Korea
6	Central Europe	Kazakhstan
7	Eastern Europe	Czech Republic
8	Eastern Europe	Serbia
9	Eastern Europe	Poland
10	Eastern Europe	Hungary
11	Europe	Estonia
12	Europe	England
13	Latin America	Cuba
14	Latin America	Peru
15	Middle East	Saudi Arabia
16	Middle East	United Arab Emirates

Source: China Government Reports.

Kenya, Japan, Indonesia, Israel, and New Zealand have also announced separate cooperation agreements with China on science, technology, and ICT.

These developments are digitalizing the Belt and Road Initiative (**BRI**) in a global connectivity context. Even before the DSR entered the lexicon, Chinese's telecom carriers (such as China Mobile and China Telecom) and telecom equipment suppliers (such as Huawei and ZTE) were already building telecommunications infrastructure, laying fiber optic tables, and integrating mobile networks for carriers around the world. They are busier than ever now under the DSR umbrella, as the global telecoms infrastructure has been overloaded with a surge in traffic due to the coronavirus pandemic and 5G network, with its larger bandwidth and low latency, is seen as a solution. Therefore, in the years ahead, China's Digital Silk Road will accelerate and expand.

DSR Greens the Belt and Road

Speaking of global connectivity, the world's largest project is China's Belt and Road Initiative (BRI) mentioned earlier in this chapter. To overcome the logistical choke points of sea routes through the straits of Malacca and the Suez Canal, the BRI, started in 2013, envisions a new silk road, reminiscent of the trade routes that connected the Tang Dynasty with the Roman and Byzantine Empires. The "Belt" refers to the Silk Road Economic Belt, an overland push that runs through Central Asia to Europe; and the "Road" is the twenty-first century Maritime Silk Road, a maritime route that runs through Southeast Asia, Africa, and Europe.

More recently, the Digital Silk Road is added to BRI as the BRI in the cyberspace (see Figure 10.2). Similar to the energy, transportation, and logistics connectivity under the BRI framework, the DSR promotes digital connectivity by teaming with BRI countries to jointly advance the construction of cross-border optical cables and other telecom networks. Moreover,

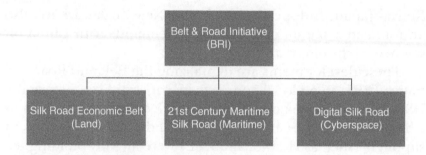

Figure 10.2   The three prongs of the BRI

it proposes to build transcontinental submarine optical cable projects, to improve spatial (satellite) information passageways, and to expand information exchanges and cooperation.

For example, Chinese telecom players have participated in the construction of a "China—ASEAN Information Harbor" in Asia, as well as "six vertical and six horizontal" key optical cables in Africa. On the satellite communication front, China has launched a collaboration with Laos and Algeria, and it has promoted collaboration with ASEAN on the BeiDou Navigation Satellite System (a competitor to GPS). Under the China-Pakistan Economic Corridor plan for Pakistan's Digital Future, the two countries have pledged to work on an upgraded fiber optic network to improve bilateral communications.

So far, digital spending along the Belt and Road still lags that on hard-infrastructure projects. As a result, the "Belt and Road" Initiative (BRI) often brings to mind a web of enormous physical infrastructures, such as roads, railways, energy pipelines, and ports. Consequently, many people think the cross-border business on the BRI will be mostly related to the abundance of surplus industrial capacity in China, much of which is heavily polluting. As such, the sustainability of the BRI is much debated globally.

What's promising, however, is that new technologies are bringing a new digital dimension to it and the related capital investments are also rising. According to the statistics from the Mercator Institute for China Studies in Berlin, by early 2020, the DSR market saw at least US$7 billion in loans and investment in cables and telecoms networks, over US$10 billion on e-commerce, mobile payments systems, and the like, and more on research and data centers. The digital dimension has expanded hugely from an initial focus on fiber-optic cables to much tech-driven 5G iABCD (IoT, AI, blockchain, cloud computing, Big Data) and "smart city" projects.

As such, the "Digital Silk Road" could potentially bring a green transformation to both infrastructure and economic models in emerging markets. Here's why:

First, the Digital Silk Road will play a constructive role in making infrastructure development more viable, efficient, and sustainable in the long run. For developing countries lacking critical infrastructures, it is highly valuable to build railroads and power plants. Adding new technologies, the DSR will help make the new infrastructure the most competitive and efficient assets possible.

For example, smart sensors and advanced monitoring systems can be integrated into infrastructure to ensure the optimization of resources. Smart grids provide a better way to match supply with demand so that power plants consume fewer fossil fuels. As an example, for the Jhang power plant in Pakistan, whose generation capacity equals the total power consumption of nearly 4 million Pakistani households, China Machinery Engineering Cooperation worked with Siemens to incorporate two high-efficiency gas turbines.

Second, as the BRI builds up land-based and maritime transportations and facilitates international trades, the DSR can integrate them with an extensive technological network to create a smart logistics system across continents. New digital technologies, such as low-cost satellites accessed by handheld smart devices, could provide real-time supply chain visibility to merchants so they know when a shipment will arrive and can plan operations in advance. This frictionless method of conducting business with a reduced focus on expanding physical infrastructures can enable more businesses and entrepreneurs to participate in global trade.

Third, the DSR narrows the life quality gap between cities and less developed areas, which makes the existing cities more sustainable in an indirect yet powerful way. As covered in earlier chapters, the biggest growth of Chinese online users, whether group e-commerce purchase or online novel reading, has come from the "small-town youths"—the young people from lower tier cities and rural areas. With the development of China's broadband infrastructure and mobile payments, small-town youths have adopted the internet lifestyle as much as those in urban areas, where the internet is in every part of daily life—from watching films and buying brands to hailing a ride and ordering food delivered to their doorstep.

When "internet lifestyle" is added to less developed regions, the small-town youths find a new path to quality livelihood at their hometown. They don't have to give up lower housing costs, shorter commute time, less traffic jams, more clean air, more direct access to fresh foods, and easier access to nature—all the things modern city dwellers are longing for. Therefore, urbanization does not have to be people migrating from rural regions, leaving them desolate, while overcrowded mega cities are challenged by transportation, environment, and social issues. As the urban–rural divide on quality of life is narrowed through digital technology and services, the global urbanization push can potentially end in more balance than conflict.

As a result, the DSR is making a direct, positive impact on individuals' life quality by the rapid construction of the information and communication technologies (ICT) network infrastructures. Take Tanzania, Africa, for example. China Telecom has helped the country to complete the construction of a key optical fiber transmission network. As a result, the level of informatization in Tanzania has been upgraded from "no internet application" to "world class" level. Thanks to the local ICTs development, the local telephone rate and internet use fees are cut into half. Tanzania's internet industries have developed as well, and the country has become one of the most important communication hubs in East Africa.

Fourth, the Digital Silk Road promotes sustainable development through the harnessing and application of Big Data to directly solve environmental challenges. For example, smart use of Big Data could enable African countries to better respond to water security issues, climate change, and natural disasters. A body named the Digital Belt and Road has reportedly set up a US$32 million program, involving experts from 19 countries and seven international organizations, to tackle natural disaster risks and natural heritage protection.

At the 2019 BRF, the BRI International Green Development Coalition was launched, which comprised more than 120 institutions that included environmental authorities, international organizations, international institutes and companies from 25 Belt and Road countries. Co-initiated by the United Nations Environment Programme (**UNEP**) and the Ministry of Ecology and Environment of China, the coalition will bring Chinese and foreign leading agencies together to conduct research and make policy recommendations on key issues (e.g. green, low-carbon, circular, and sustainable ways of production and life) to achieve the United Nations' 2030 Agenda for Sustainable Development.

Finally, the Digital Silk Road is critically important for a sustainable global economy because it helps to address one of the most fundamental challenges of the global digital economy: providing basic internet access for more than 3 billion people, roughly half the world's population, who still have no internet connectivity. While the digital economy is booming globally, the failure to narrow the Digital Divide could slow socioeconomic mobility and harden differences between the world's haves and have-nots. As the following section shows, the once "unconnected" are not merely first-time online entertainment consumers and mobile payment users. They can become digital economy entrepreneurs themselves.

Farmer Retailers, Taobao Villages, and the Alibaba University

In 2018, Alibaba and the Thai Ministry of Commerce launched the first official "Thai rice" flagship store on its Tmall marketplace. The partnership utilizes Alibaba and its logistics affiliate Cainiao's data technologies and logistics processes to optimize the cross-border flow of goods. Cainiao also works closely with the Thai Customs to promote the digitization of customs processes through technology such as Big Data and artificial intelligence, and sharing of global best practices.

As governments in Southeast Asian countries are attaching greater importance to e-commerce sectors for economic growth, Chinese companies and expertise have a greater role to play in facilitating the region to realize its potential. In China, the mobile internet creates jobs in rural areas, turning farmers into online vendors. When new channels are created to transport farm produce to cities, every farmer can be an online merchant as the demand in cities for fresh, safe agricultural products grows rapidly.

For example, fruit farmers used to sit and wait on the side of the road for buyers, but few motorists stopped to buy because most of the traffic is on the newly built motorways. Often, the farmers had to dump rotten grapes and oranges on the roadside. Now, a rural entrepreneur only needs to have a 20-square meter space, a second-hand computer, and a basic internet connection to become a global retailer. With social media-messaging-and-payment infrastructure provided by the major platforms like Alibaba and Tencent, they could easily handle large trade volume for customers from every corner of China and even reach global markets.

Taobao Village

A "Taobao Village" is defined by Alibaba as "a village in which over 10% of households runs online stores and village e-commerce revenues exceed 10 million RMB (roughly US$1.6 million) per year". These Taobao villages are the best illustration that the internet has transformed the way of life of China's rural communities.

The Taobao Village model has achieved scale to benefit about half of the total rural population in China. As of August 2019, there are a total of 4,310 Taobao Villages in 25 provinces, where 250 million out of China's total 564 million rural villagers reside, according to data from AliResearch, Alibaba Group's research arm.

Alibaba has committed providing financing, marketing support, and e-commerce training to encourage rural entrepreneurs to extend their business across the Taobao platform. For example, Alibaba group will offer financial solutions and launch special promotions on products sold by Taobao Village merchants. In addition, Alibaba also has a Taobao University (a major training division within the group) that offers seminars in rural areas so that the villagers can learn how to sell online. In some villages, long banners with large characters are seen on the walls promoting e-commerce, including "Want a better life? Get on Taobao now"!

There are even villages in which the majority of farmers are working on Taobao, Alibaba's shopping site, selling agricultural products, handicrafts, and manufactured goods, earning them the name of "Taobao Villages." According to Alibaba, the number of "Taobao Villages" was only 20 in 2013; by 2019, there were more than 4000 of them in China. (**See the "Taobao Village" box.**) A direct result of the Taobao Village boom is the creation of jobs in the most impoverished areas in China. Aside from economic prosperity, Taobao Villages have also helped to keep both parents and young workers in the village who might otherwise head to big cities to work at factories.

Furthermore, something powerful is happening on both the consumption and business sides of rural e-commerce in China. A virtuous circle is taking place whereby e-commerce is enabling more rural residents to become entrepreneurs by selling local products online, and the resultant income growth

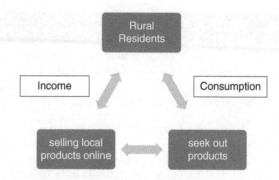

Figure 10.3   Virtuous circle of rural e-commerce

is driving up e-commerce-based consumption as hundreds of millions of villagers are being linked to e-retailing websites and becoming active customers who seek out products they can't find in their neighborhood stores (see Figure 10.3).

Across countries of the Association of Southeast Asian Nations (ASEAN), there is a high population density like in Thailand. The "China Model" of rural e-commerce may similarly empower ASEAN villagers to embrace e-commerce, maximize their earnings, and upgrade their own consumption, creating more inclusive growth across the region. These days, the "Taobao Villages" concept is popular worldwide. The annual Taobao Village Summit, where rural entrepreneurs and scholars in China share best practice for rural e-commerce businesses, is seeing an uptick in number of international participants from countries such as Rwanda, Mexico, and Malaysia, which want to replicate the model to drive economic growth at home.

For these rural entrepreneurs and similar SMEs (small and medium-sized enterprises) in emerging markets, the global marketplaces can bring in exponential growth. For example, Alibaba set up worker training programs in Malaysia to teach SMEs to sell their products on Alibaba platforms. The first overseas pilot zone of Alibaba for its eWTP (electronic world trade platform) was established in Malaysia, and by the end of March 2018, it had attracted over 2600 SMEs to operate on this

Figure 10.4   ASEAN marketplace empowered by Alibaba

platform, which opened up access for Malaysia businesses to the vast Chinese domestic market.

The synergy is best illustrated by the annual "Singles Day" shopping festival of November 11. As previously mentioned, consumers and merchants from more than 200 countries and regions participated in this 24-hour online shopping extravaganza. That is a tremendous opportunity for medium-sized and small merchants to connect with global trade markets via digital networks. During the November 11 shopping spree of 2019, organic rice grown by aboriginal villagers in Bentong of Malaysia's central Pahang state was made available on Taobao by Lazada, one of the regional leading e-commerce marketplaces (see Figure 10.4).

Founded in 2012, Lazard has business in Indonesia, Malaysia, Singapore, Thailand, Vietnam, and the Philippines. In 2016, Alibaba acquired the controlling stakes of Lazada, and since then has been trying to integrate the six markets with a complete e-commerce ecosystem, including cutting-edge technologies, logistics, and a payment system. The upgrades provide customers – Chinese and global – with better shopping experiences, while helping sellers on Lazada platform to expand their service at less cost.

The Taobao Village and Singles' Day accounts exemplify the "soft connectivity" side of the DSR (see Figure 10.5). The Digital Divide is more than the internet connectivity of individuals, the basic need of a digital economy, which requires

Figure 10.5   DSR evolves from "hard" to "soft" connectivity

physical digital infrastructure that provides "hard connectivity" for users. Furthermore, it is critical to bring to the entrepreneurs in the emerging markets an integrated digital ecosystem (the "soft connectivity") that is more robust than those local ones, so that the innovators can test and commercialize their creative ideas more effectively. In China's domestic market, connecting previously ignored small-scale entrepreneurs to the digital marketplace has led to successful startups. For emerging markets along the Digital Silk Road, the small and remotely located businesses may similarly accelerate their growth by plugging into the established online platforms of China.

The new frontier of rural e-commerce is, again, the video. As previously mentioned, with the development of fast-speed internet infrastructure in China, selling local products via livestreaming on e-commerce channels gains strong momentum for rural markets in the country. On the one hand, because new media platforms have an abundant supply of easy-to-use video tools, farmers can conveniently add videos to their marketing. On the other hand, livestreaming provides an instant and interactive way for Chinese consumers who enjoy the novelty of discovering locally grown or made products from rural villagers.

What if a Thailand farmer or Rwanda craftsman cannot be funny enough to make the cut? No worries. Captain Magellan is on the way to help. In 2020, Taobao Global, under Alibaba's Taobao Marketplace that enables overseas merchants to sell cross-border to Chinese consumers, announced the launch of its latest global initiative "Project Magellan" in Bangkok. Under Project Magellan, named after Portuguese explorer Ferdinand

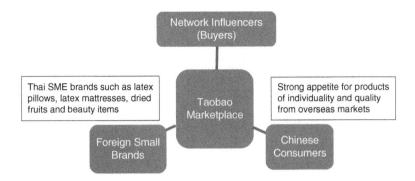

Figure 10.6   Taobao "influencers" to discover "small but beautiful" brands

Magellan who led the first expedition in history to circumnavigate the world, Taobao Global will step up its efforts to discover "small but beautiful" brands around the world and bring home products that appeal to Chinese consumers (see Figure 10.6).

Launching with Captain Magellan are the so-called "Taobao Global buyers", made up of "online celebrities" in different fields, who delve deeply into overseas lifestyles and curate quality overseas products for Chinese consumers. This platform creates business for SME brands from around the world through the videos of these influencers, who use social media posts and live streams to engage with and promote items to their followers in China and generate sales for various brands. This is the most relevant to China's young consumers who increasingly showing an appetite for products of individuality and quality. They appreciate the opportunity to interact on the video platforms with each other, key opinion leaders, and "hidden jewelry" merchants.

While the boom of cross-border e-commerce is a natural benefit of the DSR, the Silk Road in cyberspace is much more than lowering trade barriers. In the 2018 agreement between Alibaba and the Thai MOC mentioned at the beginning of this section, Alibaba, in its founder Jack Ma's words, is "committed to be a long-term partner of Thailand to help enable its digital transformation", by leveraging its strong ecosystem presence and technological know-how in e-commerce, e-payment, logistics, Big Data, and cloud services. As is discussed in the following section, the digital transformation of BRI countries is probably the most profound aspect of the DSR.

AI Anxiety and DSR Co-development

Whereas large-scale traditional infrastructure projects have so far been the focus of the BRI, the DSR is spreading internet technologies and online ecosystems that will serve as the foundation of a new digital economy for emerging markets in Africa, Southeast Asia, and elsewhere. By financing digital connectivity in underserved regions, the DSR initiative spurs economic development. Whereas the basic internet connectivity brings convenience and fun to the people in the underdeveloped regions, the link to the global online marketplaces provides them the opportunity to become online retail entrepreneurs with the access to Chinese markets.

The most profound aspect, however, is that the DSR may help BRI countries cultivating their own innovation ecosystem to meet the enormous challenges rising from the digital disruption to their existing fragile economy. With the advent of AI and new technologies, anxiety worldwide has been growing in recent years. The fear is that in the future, AI and digital tech-driven companies will have a disproportionate share of the value creation in the global value chain, and many jobs are going to be replaced by AI. No doubt, China has embraced AI and new digital economy more focused and more completely than any other nation, and its rapid transformation may also give emerging markets a sense of urgency.

China's development path during the last four decades has had four important phases (see Figure 10.7). However, because AI and digital revolution have disrupted the world economy with newly found efficiencies and productivities, the "old" China growth story, where a poorer country can, through lower labor costs and cheap resources, go for export-driven

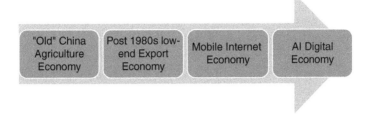

Figure 10.7   China's leap forward in four decades

businesses by manufacturing more cheaply and exporting—is over (not replicable by the current emerging markets). That means the emerging economies need to make three leaps in one, heading into the AI digital economy directly.

To stay competitive in the digital economy, all countries must embrace the same AI and tech revolution just as keenly. Yes, there are risks. And yes, there are fears and unintended consequences to technology. But the biggest risk is to miss out on the opportunity altogether. Like China, the policy makers can facilitate the digital transformation in their own regions in three important ways:

- **Clearly define digital transformation a national priority.** In China, the government has announced a sweeping vision for AI and digital economy excellence through a series of policies and initiatives.
- **Invest in digital technologies and infrastructure.** The governments can digitize government operations to stimulate consumption of digital technologies and continue to invest in expanding the information and communication technology (ICT) infrastructure to close the gap between the nations' digital haves and have-nots.
- **Promote dynamic and healthy competition to fuel innovation.** To facilitate innovation, governments should focus on building up an "innovation ecosystem" instead of purely relying on government entities and large corporations.

Through legislation, governments should ensure that entry barriers are low so that new grassroots players can compete with incumbents.

For example, governments of Middle Eastern oil-rich countries have focused on developing tech industry sectors as they are eager to diversify their economies from oil. At the forefront of this is the **United Arab Emirates (UAE)**, which is arguably the first mover in the Arab world for the technology space. Determined for the AI age, UAE has made substantial capital investments to put the fundamental, hard infrastructure in place, and in 2018, it went so far as to appoint a 27-year-old Omar bin Sultan AI Olama as the "Minister of artificial intelligence" for the state.

To cultivate domestic talents in AI, Big Data, smart cities, and related fields, UAE has overhauled its education curriculum to integrate STEM (science, technology, engineering, mathematics) to encourage young people to pursue careers in those industries. In 2019, Abu Dhabi's sovereign wealth fund Mubadala, together with Japan's Softbank Group, launched a Dh1 billion (approximately US$300 million) "global tech ecosystem" project known as "Hub 71". With attractive incentive packages, Hub71 aims to bring together seven distinct pieces of a startup ecosystem: startups, VCs, accelerators, tech companies, corporations, universities and R&D, and government— in the same fashion as China's innovation parks covered in Chapter 7.

However, the same process may impose significant challenges for many of the smaller countries, especially those that are lack of technology resources and have large labor forces that might be replaced by AI. Their biggest challenge relates to the lack of AI know-how, because the AI expertise is concentrated in only a few countries, especially the United States and China.

According to Element AI's research in May 2019, the United States, China, the United Kingdom, Australia, and

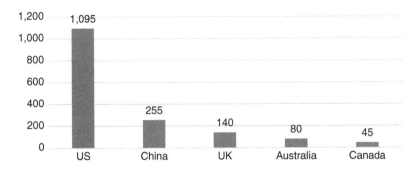

Figure 10.8   Number of high-impact AI experts (2015–2017)
Data Source: Element AI, Global AI Talent Report, 2019.

Canada lead in high-impact research in terms of number of AI experts who are publishing the most-cited papers at 21 top academic conferences. In years 2015 to 2017, these countries' total number of top researchers were the United States (1095), followed by China (255), the United Kingdom (140), Australia (80), and Canada (45), respectively (see Figure 10.8), with the United States and China way ahead of the rest of countries.

Furthermore, less-developed countries and emerging markets face an uphill battle as late-comers to the AI game. AI runs on data and that correlation leads to a self-perpetuating cycle of consolidation in industries: *the more data you have, the better your product. The better your product, the more users you gain. The more users you gain, the more data you have.* As the world's leaders in AI research, the United States and China must act to upend this trend, so that the benefits of AI revolution will be shared globally. The Digital Silk Road concept is an important start for the less-developed countries to provide the access to education, training, data, and talents that may prove to be even more valuable than financial support.

For example, in February 2018, **Nanyang Technological University, Singapore** (NTU Singapore) and Alibaba launched the Alibaba-NTU Singapore Joint Research Institute for AI Technologies. The multimillion-dollar-per-year collaboration is for an initial five years, starting with a pool of 50 researchers from both organizations.

The joint institute will seek to combine NTU's human-centered AI technology, which has been applied to areas such as health, aging, homes, and communities, with Alibaba's leading technologies, including natural language processing (NLP), computer vision, machine learning and cloud computing to explore further technology breakthroughs and real-life AI solutions. For example, an aging population is a huge issue for cities like Singapore. For that, virtual AI assistants can be deployed to improve work productivity and smart sensors can be used to watch the health of the elderly, with data stored on the cloud for continuous monitoring.

The joint institute integrates academic research with industry practices. NTU students, staff, and faculty will have opportunities to go on exchange to Alibaba's facilities and vice versa, while working on cutting-edge AI research. In addition, both parties will work toward building a crowdsourcing platform to connect researchers and industry practitioners around the world within an AI-focused R&D ecosystem, encouraging global AI experts, research institutions, and universities to join and contribute to the AI research community. The AI solutions will be tested on the NTU campus for effectiveness before Alibaba and other partners take them to the market in Singapore and rest of the world.

As with AI, cloud computing, data centers, and blockchains are critical infrastructure for all companies and organizations to operate effectively – especially tech startups. **Malaysia**, for example, is active in adopting not only AI, but also the cloud and Big Data technology. In Malaysia, SMEs and start-ups constitute more than 90% of all business establishments. Cloud computing, with its advantages of flexibility, speed to deploy, pay-as-you-use, and simplicity, is an ideal platform for the cost conscious enterprises which are often wary of upfront capital costs

Alibaba Cloud, the cloud computing arm of Alibaba Group, in 2017, started its operations of Malaysia's first global public cloud platform. According to a statement from the company, the Malaysian enterprises will be able to access cost-effective, cloud-based services to accelerate and scale at a much higher rate. The Alibaba Cloud services would likely be most useful to governments and large corporations that have huge data needs, but the modularity of the data center's products will also help small businesses.

Indonesia is another ASEAN country that Alibaba Cloud has aggressively expanded into. In 2018, it commenced its data center operation in Indonesia, which is part of Alibaba Cloud's effort to support the Indonesian government's initiative to create 1000 startups by 2020. That data center is also the first global public cloud platform in the country. By migrating their IT infrastructure to Alibaba Cloud, the SMEs and startups can leverage affordable cloud services to scale rapidly, accelerate innovation, and reduce cost. Subsequently, they can expand their geographic reach through Alibaba Cloud's established global network to support international operations and better compete in global trade.

By partnering with other cloud services companies, Alibaba Cloud had more than 20 overseas data centers outside of China by 2020. (In Indonesia, it had opened its second data center.) In July 2018, Alibaba confirmed it was in talks with BT to partner in the United Kingdom to offer its cloud services. The goal of this deal would be to allow Alibaba to compete with Amazon in the United Kingdom. In South Korea, Alibaba has a partnership with SK Group to provide cloud services.

The latest addition to the tech mix is blockchain. China, in April 2020, launched the Blockchain Services Network (BSN), which is a critical part of China's national blockchain strategy that was announced by President Xi in late November 2019. The BSN is an ambitious effort to include as many blockchain frameworks as possible and make them accessible under one uniform standard on the BSN platform. As such, it's the largest

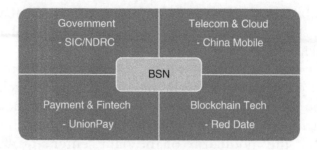

Figure 10.9   The open blockchain-based service network, BSN

blockchain ecosystem in China, and in the DSR context, it is rapidly expanding its network overseas.

The open BSN was jointly developed by central government, state corporations, and private tech companies (see Figure 10.9), including Chinese government policy think tank the State Information Center (SIC, which is under the NDRC—National Development and Reform Commission—which also set up the AI national development policy discussed in Chapter 1), and China's state-run telecom giant China Mobile, and the Chinese government-supported payment card network China UnionPay (see more background relating to fintech discussions in Chapter 5), and the private tech company Red Date Technology.

According to its official white paper, BSN is a cross-cloud, cross-portal, cross-framework global infrastructure network used to deploy and operate all types of blockchain applications. Previously, each participant in a traditional consortium chain application must build and operate its own exclusive node and respective consensus mechanism. Each node needs to use a physical server or cloud service to connect with one another through the internet or an internal network, thereby forming an isolated blockchain application similar to a local-area network.

In this isolated blockchain structure, the application designer needs to establish a new blockchain operating environment for each consortium chain, which is highly costly. For example, to deploy blockchain applications at the platforms of major cloud providers (such as Alibaba, Huawei) may cost the users tens of thousands of dollars a year. Worse, the server resources are often not fully used. According to Red Date, the tech firm behind the BSN, only 2–3% of enterprise users would need more than 1000 transactions per second (TPS), which allows users to make full use of the cloud services.

Therefore, the high cost brought by deployment and maintenance is a major barrier for blockchain entrepreneurs. **First**, the BSN launch will allow companies to access ultra-low cost blockchain cloud computing services because it enables customers to have the services in much smaller units. Users can pay for exactly how much they need. Target pricing is less than US$400 per year, which would allow any SME or individual access to the critical tools to participate in the digital economy and drive adoption and financial inclusion opportunities.

Second, the BSN aims to simplify blockchain application developments. It provides a public blockchain resource environment for developers with the concept of internet. Just like building a simple website on the internet, developers can deploy and operate blockchain and distributed ledger applications conveniently and at low cost. (Distributed ledger technology, or "DLT", is the technology behind blockchain.) According to Red Date, building an application for 80% of the BSN users would be as easy as filling a form online. ("They don't even have to write their own smart contracts; all they need to do is select one of them in our system".)

Third, the BSN plans to enhance the connectivity between different blockchain applications (see Figure 10.10). In the traditional isolated blockchain context, various applications of different technical standards cannot be unified, thus business data is unable to interact, which restricts the broad adoption of blockchain technology. The BSN white paper indicates developers can use a single private key to deploy and manage applications on multiple frameworks, and, at the same time,

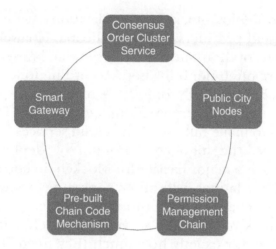

Figure 10.10   Five major parts of the BSN framework

realize interconnectivity and mutual communication. (Within this process, each framework retains the unique features of its own smart contract and consensus mechanism.)

In summary, the BSN is set to include as many blockchain frameworks as possible and make them accessible under one uniform standard so that it could be deployed nationally and globally at low cost. In China, all city governments, state-owned enterprises, and IT framework operators are gearing up to adopt and interoperate with the protocol. Its multi-cloud architecture has already included China Unicom, China Telecom, China Mobile, and Baidu Cloud from the domestic sector as well as Amazon AWS, Microsoft Azure, and Google Cloud of the West.

At the April 2020 launch, the Blockchain Service Network Development Alliance stated that the network had 128 public nodes. China had 76 of these nodes already in the network at that time with 44 under construction, and the remaining 8 overseas city nodes (including Singapore, which already tested BSN during its pilot phase), which covered six continents. By the end of 2020, the BSN plans to have at least 200 nodes in the network. Therefore, right at its inception and within China alone, the BSN ecosystem was instantly the largest blockchain ecosystem in the world. Along the DSR, more BSN nodes probably would be added to the network, as institutions and governments of BRI countries may deploy and adopt blockchain at this convenient, low-cost platform.

Israel and Everyone: Caught in the Middle

As seen in the previous sections, China's expanding "Digital Silk Road" into ASEAN, Africa, and the Middle East has immense positive implications for their growing needs for both digital and physical infrastructure and connectivity. Setting aside the usual debates about democracy, privacy, administrative transparency, interest alignment between stakeholders—all of which are legitimate concerns deserving their own analysis—those markets could substantially benefit from partnering with China for their own version of digital transformation, such as the "Thailand 4.0 policy initiative", the "Smart Nation initiative" of Singapore, and the "Digital Transformation Strategy for Africa".

However, the DSR is now caught up in the broader United States–China tech confrontation. The 5G network, for example, is one of the frontline battles. The United States and some partner countries have been pushing to exclude Chinese 5G equipment makers from their networks. They believe that greater involvement by Chinese companies in multilateral technology standards-setting could materially alter the course of global norms in ways the United States and other democracies could not support. Meanwhile, the infrastructure financing and trade relationship from China is really attractive to the BRI countries keen to upgrade their digital economy ecosystem.

The reality is that China, due to its sheer size, is the largest trading partner of most other Asian countries, including every treaty ally of the United States in the region, as well as many more around the world. As such, many countries want to cultivate good relations with China and to participate in its economic growth—especially its digital revolution. Interestingly, the growing United States–China technology tensions have led to unexpected opportunities for third-party countries—but probably not for too long.

Take **Israel**'s flourishing semiconductor industry, for example. Israel is a tech powerhouse, particularly in cybersecurity and weapons technology. Among foreign companies listed on NASDAQ, Israeli companies trail only Chinese companies. When Silicon Valley started closing its doors to Chinese money, Israel, whose innovative tech sector inspired the nickname "Startup Nation", became an alluring destination: China gets a valuable pipeline of technology, and Israel teams up with a deep-pocketed backer.

Since the 1980s, Israel has carefully walked the US–China technology tensions tightrope, trying to balance its commercial and security interests with the two great powers. What Israel strove to form is a technology triangle with China and the United States, enabled by maintaining well-thought-out controls on dual-use technology transfers (see Figure 10.11). As a workaround to US–China trade tariffs, a US semiconductor giant increased direct sales from its Israeli plant to Chinese buyers, thus boosting semiconductor trade between Israel and China by 80% in 2018. In parallel, China increased investments

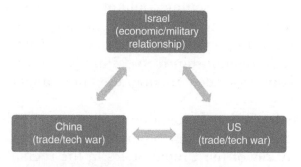

Figure 10.11   Israel balancing United States–China relationships

in Israeli semiconductor companies, driving up demand for Israeli-designed chips.

However, the importance of chips for technological advancement and military use mean Israel's position in this technology triangle is bound to attract Washington's attention. The growing dual-use nature (meaning they have both military and commercial applications, such as drones and artificial intelligence) of technology threatens to over-throw Israel's careful efforts to expand trade with Beijing, while avoiding the sales of security technologies that would increase Chinese military capabilities and anger Washington. Israeli entrepreneurs are also fearful that receiving Chinese investments could negatively affect business in the United States.

Pressured by a Trump administration that worries about Israel's dalliance with China, Israel is moving to create an inter-agency government body—similar to the US' CFIUS—to review foreign investments in sensitive areas of its economy. Once set up, the regulator reportedly would focus on dual-use products that have military and commercial applications, such as semi-conductors, drones, and artificial intelligence. If tensions escalate further, Israel could risk finding its American and Chinese trade partners less forgiving of its delicate balance.

Nevertheless, Israel is still trying hard to maintain a balance. For example, relating to 5G networks, Israeli government pushed back against the United States' position on completely banning Huawei due to commercial interests, while it also avoids using Huawei's equipment in its critical infrastructure for security concerns. Many countries, just like Israel, do not wish to be forced to choose between the United States and China. They also hope that the United States understands that if other countries develop ties with China, that does not necessarily mean that they are working against America. (And of course, these countries hope for the same understanding from China, too, if they strengthen their ties with the United States.)

However, more and more countries would find, that a hard balancing act is bound to become even harder as the China–US political and security tensions continue to escalate. This is a concern to all the countries, as they are afraid of being forced into invidious choices between the two superpowers. As such, the underlying rationale of the DSR and digital economy—that is, global connectivity and collective development, which is even more important than the DSR infrastructure projects themselves—is at risk.

Innovative, Invigorated, Interconnected, and Inclusive

As this book concludes, there is no doubt that China is a tech superpower; its digital globalization is just getting started and its rise in AI and digital technologies will have global economic, social, and geopolitical implications that will resonate for decades. Today, the US internet giants of GAFA (Google, Apple, Facebook, Amazon) are the global leaders in innovative technology; however, the next generation of startups in emerging markets may look to the likes of Alibaba and Tencent for inspiration.

China's increasing prominence on the world's digital stage also means that the country can contribute, and even lead, broader debates on global governance issues such as digital sovereignty, cross-border data flows, and cybersecurity. There is intense debate around the world about how to react to and govern the digital world. While there are clear rules governing global trade in traditional goods under the framework of the WTO (which is also challenged by the recent de-globalization trend), the digital economy still lacks a well-defined governance framework or widely accepted regulations on the data underneath.

As China pushes the Digital Silk Road Initiative forward and calls other nations to create "a community of common destiny in cyberspace", it is in all companies, nations, and other global stakeholders' interests to ensure that they engage with China and its digital economy on four key areas of global digital governance:

- **Develop digital (data) regulations**. The digital revolution has leaped ahead of existing legislations, and all nations have merely nascent data protection and Internet management policies. The European General Data Protection Regulation (GDPR) has provided a "framework" for the European market, but there is no universal standard for data rights, privacy, and management.
- **Rebuild the intellectual property (IP) system**. Digital revolution has challenged the existing patent and copyright laws, and the whole IP system requires a substantial reform in the digital age.
- **Manage cybersecurity challenges**. Cybersecurity is a complex transnational issue for all nations, which requires global cooperation for ensuring a safe Internet. There exists huge potential for seeking common ground in raising cyber defense and resilience against hackers.
- **Build an open digital economy**. The geopolitical tensions and technology competition has highlighted the risk that the world evolves into a period of divergence following globalization. Nations should stay committed to globalization for the digital economy to continue flourishing.

However, 2019–2020 may be remembered as an important inflection point for the global digital economy, from collaborative co-existence to head-on tension, as illustrated by the tech war between the United States and China, the two digital superpowers of the world. The tech war could derail next-generation innovation and fragment what has become an increasingly global digital community. For one, the development of technology would become more expensive, if the global supply chain is disrupted. For another, local regulations might force companies to develop two different sets of technologies, where China uses Chinese products and America uses American products, and consumers will have problems of compatibility.

That could mean that users may end up with multiple smartphones—one that works in China and another that works outside the country, and vice versa. For international travelers who have grown accustomed to being able to jump on wireless networks while abroad, they may find their connectivity to the internet more frequently cut off. More seriously, enterprises may see the end of global software that has allowed for cross-Pacific communications and research projects, such as shared development of artificial intelligence technologies.

Therefore, although different global stakeholders may use terms different from the Digital Silk Road (e.g. "Shared Digital Future" for the World Economic Forum), all players agree that a global dialogue on openness and mutual benefit should happen in a new context (like the WTO framework for global trade). The new global "norms" are critical for the free cross-border flow of trade, capital, data, and talent to continue, such that digital economy innovation could continue to serve as the growth engine for the world.

As such, "Data Free Flow with Trust" (DFFT), initiated by Japan at the June 2019 Osaka G20 summit, emerged as a rare bright spot for potential harmonization at a time when national views on data and privacy vary across the globe and barriers to cross-border data flows, which confine data within a country's borders ("data localization"), are on the rise. (**See the "DFFT on Osaka Track" box.**) The Japanese government's initiative to find consensus on global data governance may seem overly ambitious, but to simply start the global conversation is a critical step to revive multilateral cooperation in the fractured tech world.

DFFT on Osaka Track

At the special meeting on the digital economy during the 2019 summer Osaka G20 summit, sitting between U.S. President Donald Trump on his right and Chinese President Xi Jinping on his left, Prime Minister Abe formally declared the launch of the "Osaka Track" under the roof of the World Trade Organization (WTO).

The Osaka track is an overarching framework promoting cross-border data flow with enhanced protections—the concept of "Data Free Flow with Trust" (DFFT), and the WTO will also be reformed accordingly to be relevant again as a guardian of free and fair international trade. To some extent, the DFFT could make the WTO remain relevant in the age of digital economy.

The key principle for DFFT data governance is to standardize rules in global movement of data flows: a world in which data crosses national borders freely among countries with high levels of privacy protection, data security and intellectual property rights. As such, DFFT also suggests that businesses should not be required to use or locate computing facilities in a country as a condition for doing business there ("data localization"). After the meeting, 24 countries formally signed a statement affirming the DFFT Osaka Track concept in general terms.

The DFFT Osaka Track could well be Minister Abe's most important initiative ever, as he said earlier at his 2019 Davos speech, "I would like Osaka G-20 to be remembered as the summit that started worldwide data governance".

Of course, the DSR, DFFT and many similar multinational and multistakeholder initiatives must collaborate to achieve the common goal. That is far from an easy task. The cross-border cyberspace governance debate—rules for who controls data, and therefore, harnesses their value and increase the nation's innovation power—is not merely a technical problem; instead, it is at the core of the geopolitical competition that will shape the twenty-first century. To add to the complexity are the deep divides on the topic, not only between China and the United States, but also among governments across the European Union, Japan, India, and the emerging markets that have philosophical differences on how they approach the data economy issues.

Therefore, it is critical for the United States and China, the two tech superpowers, to reach a new equilibrium to collectively lead the future 5G iABCD innovation with the rest of the world; because, at the core of digital economy is the free flow of trade, capital, intellectual, and data. Our global economy is interconnected and there is no turning back. As the G20 leadership together declared at the G20 Summit 2016 in Hangzhou, China: collectively, the digital economy revolution will build an innovative, invigorated, interconnected, and inclusive world economy.

Bibliography

Chapter 1

Berthiaume, Dan. "Alibaba reveals secret to its Singles Day mega-sale success", Chain Store Age, 20 November 2019. https://chainstoreage. com/alibaba-reveals-secret-its-singles-day-mega-sale-success

Byers, Dylan. "TikTok sues over Trump executive order that would ban the app in the U.S.", NBC news, 24 August 2020. https://www.nbcnews. com/tech/tech-news/tiktok-sues-over-trump-executive-order-would-ban-app-u-n1237859

Carman, Ashley. "Why Zoom became so popular", The Verge, 3 April 2020. https://www.theverge.com/2020/4/3/21207053/zoom-video-conferencing-security-privacy-risk-popularity

CBN Editor. "Ant Financial rebrands itself as 'Ant Technology'", China Banking News, 26 June 2020. http://www.chinabankingnews.com/ 2020/06/26/ant-financial-rebrands-itself-as-ant-technology/

Chen, Lulu Yilun, and Vinicy Chan, Katie Roof, and Zheping Huang. "TikTok owner's value exceeds $100 billion in private markets", BloombergQuint, 20 May 2020. https://www.bloombergquint.com/ business/tiktok-owner-s-value-surpasses-100-billion-in-private-markets

cnTechPost. "IDC releases top 10 predictions for China cloud computing market in 2020", cnTechPost, 19 February 2020. https://cntechpost. com/2020/02/19/idc-releases-top-10-predictions-for-china-cloud-computing-market-in-2020/

Davidson, Helen. "China starts major trial of state-run digital currency", The Guardian, 28 April 2020. https://www.theguardian.com/world/2020/ apr/28/china-starts-major-trial-of-state-run-digital-currency

Economist. "The digital side of the Belt and Road Initiative is growing", The Economist, 6 February 2020. https://www.economist.com/special-report/2020/02/06/the-digital-side-of-the-belt-and-road-initiative-is-growing

Feng, Coco, and Minghe Hu. "How a decade of smartphone apps changed the way people live, work and play in China", SCMP, 30 December 2019. https://scmp.com/tech/apps-social/article/3043688/how-decade-smartphone-apps-changed-way-people-live-work-and-play

Finnegan, Conor. "US looking at banning Chinese social media app TikTok as security threat", ABC News, 7 July 2020 https://abcnews. go.com/Politics/us-banning-chinese-social-media-app-tiktok-security/ story?id=71647269

Galvan, Bryan, and Kelly Le. "China's new DCEP could fast-forward the nation into a cashless society", Forkast, 11 June 2020. https://forkast. news/china-dcep-accelerate-nation-cashless-society-privacy-digital-currency/

Gu, Shengzu, "Digital economy a new growth engine", China Daily, 5 June 2020. https://global.chinadaily.com.cn/a/202006/05/ WS5ed97e71a310a8b24115af95.html

Hellard, Bobby. "Who has banned Zoom and why?" CloudPro, 9 April 2020. https://www.cloudpro.co.uk/collaboration/8518/who-has-banned-zoom-and-why

Harris, Mark. "Google and Facebook turn their backs on undersea cable to China", TechCrunch, 6 February 2020. https://techcrunch. com/2020/02/06/google-and-facebook-turn-their-backs-on-undersea-cable-to-china/

Huang, Zheping. "Xiongan calls in ConsenSys to bring blockchain technology to Xi Jinping's dream city", SCMP, 23 July 2018. https://www.scmp. com/tech/article/2156396/xiongan-calls-us-company-consensys-bring-blockchain-technology-xi-jinpings

Jakobson, Leo. "Walmart China turns to blockchain to reassure consumers wary of food safety", Modern Consensus, 26 June 2019. https:// modernconsensus.com/technology/walmart-china-turns-to-blockchain-to-assure-consumers-wary-of-food-safety/

Kelly, Samantha. "Zoom's massive 'overnight success' actually took nine years", CNN Business, 27 March 2020. https://www.cnn. com/2020/03/27/tech/zoom-app-coronavirus/index.html

Kharpal, Arjun. "With Xi's backing, China looks to become a world leader in blockchain as US policy is absent", CNBC, 15 December 2019. https:// www.cnbc.com/2019/12/16/china-looks-to-become-blockchain-world-leader-with-xi-jinping-backing.html

Kitson, Andrew, and Kenny Liew. "China Doubles Down on Its Digital Silk Road", CSIS, Nov. 14, 2019. https://reconnectingasia.csis.org/analysis/ entries/china-doubles-down-its-digital-silk-road/

Lin, Liza, Jing Yang, and Eva Xiao. "Microsoft's Talks to Buy TikTok's US Operations Raise Ire in China", Wall Street Journal, 3 August 2020. https://www.wsj.com/articles/microsofts-talks-to-buy-tiktoks-u-s-operations-raise-concerns-in-china-11596465664?mod=hp_lead_pos2

Ma, Winston. "Why the internet is yesterday's news in China's digital leap", World Economic Forum, September 2019. https://www.weforum.org/agenda/2019/09/why-the-internet-is-yesterdays-news-in-chinas-digital-leap-forward

McKinnon, John, and Lingling Wei. "Corporate America worries WeChat ban could be bad for business", Wall Street Journal, Aug. 13, 2020. https://www.wsj.com/articles/corporate-america-worries-wechat-ban-could-be-bad-for-business-11597311003?mod=hp_lead_pos1

McMorrow, Ryan, and Nian Liu. "Xi Jinping's endorsement of blockchain sparks China stocks frenzy", Financial Times, 28 October 2019. https://www.ft.com/content/2789d21a-f955-11e9-98fd-4d6c20050229

Musil, Steven. "WeChat users sue Trump administration over app ban", CNET, 23 August 2020. https://www.cnet.com/news/wechat-users-sue-trump-administration-over-app-ban/

Ouyang, Shijia, and Chen Jia. "China embracing blockchain technology as new frontier of innovation", China Daily, 1 November 2019. https://www.chinadaily.com.cn/a/201911/01/WS5dbb96efa310cf3e35574e50.html

Pan, Che. "Didi partners with China's central bank on digital currency research and development", SCMP, 8 July 2020. https://www.scmp.com/tech/apps-social/article/3092379/didi-partners-chinas-central-bank-digital-currency-research-and

Rapoza, Kenneth. "China's digital currency to be given a test drive by U.S. companies", Forbes, 23 April 2020. https://www.forbes.com/sites/kenrapoza/2020/04/23/chinas-digital-currency-to-be-given-a-test-drive-by-us-companies/#2d214aa87992

Satariano, Adam, and Stephen Castle. "U.K. bans Huawei from 5G network, raising tensions with China", New York Times, July 14, 2020. https://www.nytimes.com/2020/07/14/business/uk-bans-huawei-from-5g-network-raising-tensions-with-china.html

Shepardson, David and Eric Beech. "Trump orders ByteDance to divest interest in U.S. TikTok operations within 90 days", Reuters, 14 August 2020. https://www.reuters.com/article/us-usa-tiktok/trump-orders-bytedance-to-divest-interest-in-u-s-tiktok-operations-within-90-days-idUSKCN25B00K

Shieber, Jonathan. "West Summit looks beyond CIC for newest fund", Private Equity News, 25 June 2012. https://www.penews.com/articles/westsummit-looks-beyond-cic-20120625

Technology News China Friday, "China's trillion-dollar campaign fuels a tech race with the U.S.", China Technology News, 12 June 2020. http://www.technologynewschina.com/2020/06/chinas-trillion-dollar-campaign-fuels.html?m=1

United States Department of Justice. "Attorney General William Barr delivers the keynote address at the Department of Justice's China Initiative Conference", 6 February 2020. https://www.justice.gov/opa/speech/attorney-general-william-p-barr-delivers-keynote-address-department-justices-china

United States Department of Justice, "Team telecom recommends that the FCC deny Pacific Light Cable Network System's Hong Kong undersea cable connection to the United States",17 June 2020. https://www.justice.gov/opa/pr/team-telecom-recommends-fcc-deny-pacific-light-cable-network-system-s-hong-kong-undersea

United States Treasury. "CFIUS Overview". https://home.treasury.gov/policy-issues/international/the-committee-on-foreign-investment-in-the-united-states-cfius/cfius-overview

Wa, Coonie. "Xiongan new area master plan underscores blockchains' Opportunity", Zilian8, 3 April 2020. https://en.zilian8.com/1005.html

Yi, Ding. "Alibaba-owned livestreaming platform Taobao Live sees rapid growth amid outbreak", Caixin Global, 20 February 2020. https://www.caixinglobal.com/2020-02-20/alibaba-owned-livestreaming-platform-taobao-live-sees-rapid-growth-amid-outbreak-101518181.html

Zhang, Jane. "China's AI champions are already powering a mind-boggling array of processes and this will rise in 2020", SCMP, 1 January 2020. https://scmp.com/tech/start-ups/article/3044188/chinas-ai-champions-are-already-powering-mind-boggling-array

Chapter 2

Alizila. "The rise of male beauty in China", Alizila, 27 March 2019. https://www.alizila.com/the-rise-of-male-beauty-in-china/

BBC. "Didi Chuxing: Apple-backed firm aims for one million robotaxis", 23 June 2020. https://www.bbc.com/news/technology-53157368

Brennan, Tom, and Susan Wang. "Cyber celebrities monetize their fame with Taobao", Alizila, 14 June 2016. https://www.alizila.com/cyber-celebrities-take-to-taobao-to-monetize-their-fame/

Cheng, Evelyn. "Singles Day sales hit a record high as Chinese buyers rack up their credit card bills", CNBC, 15 November 2019. https://www.cnbc.com/2019/11/15/singles-day-sales-hit-record-high-as-chinese-buyers-rack-up-credit-card-bills.html

Huy, Quy. "For Alibaba, Singles' Day is about more than huge sales", Harvard Business Review, 11 December 2019. https://hbr.org/2019/12/for-alibaba-singles-day-is-about-more-than-huge-sales

Liang, Meng, and Han Su. "'Planting Grass' and RED: How e-commerce in China is interweaved with social media", Global Media, 1 December 2019. http://globalmedia.mit.edu/2019/12/01/2058/

McCourt, David. "Happy birthday Xiaomi! The milestones along its meteoric rise", NextPit, April 2020. https://www.nextpit.com/happy-birthday-xiaomi-10-years

Ni, Vincent, and Yitsing Wang. "Coronavirus: Can live-streaming save China's economy?" BBC World Service, 6 May 2020. https://www.bbc.com/news/business-52449498

Ouyang, Iris. "Coronavirus: Live-streaming sales prove a lifeline for China's small retailers as pandemic disrupts business models and consumer behaviour", SCMP, 25 April 2020. https://www.scmp.com/business/article/3081454/live-streaming-sales-prove-lifeline-chinas-small-retailers-coronavirus

Qu, Tracy. "A red-hot Chinese shopping-review app shows the future of your online shopping experience", Quartz, 28 June 2019. https://qz.com/1634577/chinas-xiaohongshu-shows-the-future-of-your-social-shopping-experience/

Qu, Tracy. "Baidu tops Microsoft and Google in teaching AI to understand human language, thanks to differences between Chinese and English", SCMP, 30 December 2019. https://scmp.com/tech/big-tech/article/3043895/baidu-tops-microsoft-and-google-teaching-ai-understand-human-language

SCMP. "Baidu CEO Robin Li has 'more confidence' for next year after company reports better-than-expected third quarter results on growth in video, cloud services", SCMP, 7 November 2019. https://scmp.com/tech/enterprises/article/3036671/baidu-ceo-robin-li-has-more-confidence-next-year-after-company

Technology Magazine. "Ride-hailing firm Didi Chuxing (DiDi) has experienced a meteoric rise since its founding in 2012, with the company expanding into autonomous vehicles", Technology Magazine, 5 June 2020. https://www.technologymagazine.com/ai/how-didi-leveraging-ai-autonomous-robotaxis

SCMP. "How Alibaba powered billions of transactions on Singles' Day with 'zero downtime'", SCMP, 20 November 2019. https://scmp.com/tech/e-commerce/article/3038539/how-alibaba-powered-billions-transactions-singles-day-zero-downtime

Xie, Stella. "A $7 credit limit: Jack Ma's Ant lures hundreds of millions of borrowers", Wall Street Journal, 8 December, 2019. https://www.wsj.com/articles/a-7-credit-limit-jack-mas-ant-lures-hundreds-of-millions-of-borrowers-11575811989

Zhang, Jane, Sarah Dai, and Coco Feng. "China's record Singles' Day offers glimpse of future shopping trends as buyers embrace live streaming", SCMP, 13 November 2019. https://scmp.com/tech/e-commerce/article/3037401/chinas-record-singles-day-offers-glimpse-future-shopping-trends

Chapter 3

Brzeski, Patrick. "China box office: Slow growth marks 'new normal'", Hollywood Reporter, 8 January 2020. https://www.hollywoodreporter.com/news/china-box-office-slow-growth-marks-new-normal-1267762

China Banking News. "Guangzhou to launch China's first blockchain and Big Data-driven food monitoring platform", China Banking News, 25 March 2020. http://www.chinabankingnews.com/2020/03/25/guangzhou-to-launch-chinas-first-blockchain-and-big-data-driven-food-monitoring-platform/

Davis, Rebecca. "China's box office hit new heights in 2019, as Hollywood's share shrank", Variety, 2020. https://variety.com/2020/film/news/china-box-office-2019-review-ne-zha-wandering-earth-avengers-1203455038/

Global Times. "Kids develop 'big-head' disease after using solid drink sold as milk powder", Global Times, 13 May 2020. https://www.globaltimes.cn/content/1188234.shtml

Huang, Echo. "Blockchain will track how meat gets from Australian farms to Chinese tables", Quartz, 7 March 2018. https://qz.com/1223228/jd-is-using-blockchain-to-track-how-meat-gets-from-australian-farms-to-chinese-tables/

JD press release. "Walmart, JD.com, IBM, and Tsinghua University launch a blockchain food safety alliance", 14 December 2017. https://jdcorporateblog.com/walmart-jd-com-ibm-and-tsinghua-university-launch-a-blockchain-food-safety-alliance-in-china/

Lang, Brent. "Sluggish China box office brings down 2016's global totals", Variety, 2017. https://variety.com/2017/film/box-office/2016-global-box-office-1201968877/

Levin, Dan, and Crystal Tse. "In China, stomachs turn at news of 40-year-old meat peddled by traders", New York Times, 24 June 2015. https://www.nytimes.com/2015/06/25/world/in-china-stomachs-turn-at-news-of-traders-peddling-40-year-old-meat.html?referringSource=articleShare

Liao, Rita. "China's top short video apps and e-commerce giants pally up", TechCrunch, 27 May 2020. https://techcrunch.com/2020/05/27/chinas-top-short-video-apps-and-e-commerce-giants-pally-up/

Pinduoduo Press Release. "Pinduoduo subscribes to US$200 million in GOME convertible bonds in strategic partnership consumer-to-manufacturer (C2M)", 19 April 2020. http://investor.pinduoduo.com/news-releases/news-release-details/pinduoduo-subscribes-us200-million-gome-convertible-bonds

Qu, Tracy. "Live in the time of coronavirus: Sleeping man, farmers ride wave of streaming popularity in China", SCMP, 19 February 2020. https://www.scmp.com/print/tech/apps-social/article/3051394/live-time-coronavirus-sleeping-man-rural-farmers-ride-wave

Wang, Jing. "New Freshippo store formats cater to different consumer needs", Alizila, 3 March 2020. https://www.alizila.com/freshippo-store-formats-cater-to-different-consumer-needs/

Chapter 4

Bursztynsky, Jessica. "TikTok nabs Disney's streaming boss to be its new CEO", CNBC, 18 May 2020. https://www.cnbc.com/2020/05/18/tiktok-nabs-disneys-streaming-boss-to-be-its-new-ceo.html

Chan, Connie. "Outgrowing advertising: Multimodal business models as a product strategy", A16Z, 7 December 2018. https://a16z.com/2018/12/07/when-advertising-isnt-enough-multimodal-business-models-product-strategy/

Davis, Rebecca. "iQIYI makes strides in interactive video", Variety, 2019. https://variety.com/2019/film/news/iqiyi-interactive-video-his-smile-1203249952/

Feng, Coco. "Short video is now more attractive than news sites for online advertisers in China", SCMP 15 January 2020. https://scmp.com/tech/e-commerce/article/3046156/short-video-now-more-attractive-news-sites-online-advertisers-china

Feng, Coco, and Minghe Hu, and Tracy Qu. "TikTok overtakes Facebook to become world's second most downloaded app as rivals challenge its short video dominance", SCMP, 15 January 2020. https://scmp.com/tech/apps-social/article/3046039/tiktok-overtakes-facebook-become-worlds-second-most-downloaded-app

Frater, Patrick. "Sony paying $400 million for stake in Bilibili, Chinese online platform", Variety, 2020. https://variety.com/2020/biz/asia/sony-buys-stake-in-china-bilibili-1234575663/

He, Laura, and Karen Yeung. "Tencent's China Literature most-profitable IPO debut in a decade after value soars as much as 100 percent", SCMP, 8 November 2017. https://www.scmp.com/business/companies/article/2118854/china-literature-shares-soar-63-cent-hong-kong-trading-debut iQIYI press release. "iQIYI launches interactive video platform plug-in in Premiere Pro, an easier and more accessible way of making interactive videos", 23 July 2019. https://www.iqiyi.com/common/20190724/beae335f764e176e.html

Kingdom, Magpie. "Bilibili started out as a platform for Japanese anime: It's now the center of China's social video boom", Splice Media, 11 January 2018. https://www.splicemedia.com/china-bilibili-platform-social-video-boom/

Klein, Jodi. "US Senate panel unanimously approves ban on TikTok on government devices", SCMP, 23 July 2020. https://www.scmp.com/tech/article/3094318/us-senate-panel-unanimously-approves-ban-tiktok-government-devices

Lee, Emma. "Alibaba removes youngest partner Jiang Fan after affair accusation", Technode, 27 April 2020. https://technode.com/2020/04/27/alibaba-removes-youngest-partner-jiang-fan-after-affair-goes-public/

Linder, Alex. "Alibaba senior exec demoted over alleged affair with online celeb", Shanghaiist, 29 April 2020. http://shanghaiist.com/2020/04/29/alibaba-senior-exec-demoted-over-alleged-affair-with-online-celeb/

Qu, Tracy. "I took a class on how to get rich on Douyin, the Chinese TikTok, and learnt that originality is overrated", SCMP, 14 January 2020. https://scmp.com/tech/apps-social/article/3045854/i-took-class-how-get-rich-douyin-chinese-tiktok-and-learnt

Qu, Tracy, and Iris Deng. "Streaming video operator Bilibili at a crossroads with focus on increasing users", SCMP, 12 November 2019. https://scmp.com/tech/apps-social/article/3037193/streaming-video-operator-bilibili-crossroads-planned-revamp

Rapoza, Kenneth. "After India, U.S. considers banning TikTok for real this time", Forbes, 7 July 2020. https://www-forbes-com.cdn.ampproject.org/c/s/www.forbes.com/sites/kenrapoza/2020/07/07/after-india-us-considers-banning-tiktok-for-real-this-time/amp/

Rapp, Jessica. "Targeting China's Generation Z? The cool kids are on Bilibili", Parklu, 26 June 2019. https://www.parklu.com/chinas-generation-z-bilibili/

Shen, Xinmei. "Netflix has interactive movies and China's streaming sites want them, too", Abacus, 21 August 2019. https://www.abacusnews.com/digital-life/netflix-has-interactive-movies-and-chinas-streaming-sites-want-them-too/article/3023700

Xu, Jieru. "How do China's internet celebrities differ from America's?" Ruggles Media 18 September 2018. https://camd.northeastern.edu/rugglesmedia/2018/01/27/internet-celebrity-and-diversity/

Wang, Yue. "iQiyi is no longer content with being the Netflix of China", Forbes, 29 May 2019. https://www.forbes.com/sites/ywang/2019/05/29/iqiyi-is-no-longer-content-with-being-the-netflix-of-china/#7a1a05c829cb

Chapter 5

Alipay. "Alipay's AI wealth tech platform adopted by China's asset managers", Finextra, 20 June 2019. https://www.finextra.com/pressarticle/78864/alipays-ai-wealth-tech-platform-adopted-by-chinas-asset-managers

Bloomberg News. "China's first AI fund learned from the country's best traders", 24 July 2019. https://www.bloomberg.com/news/articles/2019-07-24/china-s-first-ai-fund-learned-from-the-country-s-best-traders

Borak, Masha. "Debt: The secret sauce of Alibaba's Singles' Day success", Technode, 22 November 2017. Singles' Day https://technode.com/2017/11/22/huabei-singles-day/

Cheng, Evelyn. "China taps blockchain technology to boost financing for businesses hit by virus", CNBC, 15 February 2020. https://www.cnbc.com/2020/02/15/coronavirus-china-taps-blockchain-tech-to-help-firms-hit-by-virus.html

China Banking News. "Chinese banks emerge as leading blockchain adopters", 30 March 2019. http://www.chinabankingnews.com/2019/03/30/chinese-banks-emerge-as-leading-blockchain-adopters/

China Banking News. "Agricultural Bank of China teams up with online supply chain finance platform on fintech innovation", 9 May 2019. http://www.chinabankingnews.com/2019/05/09/agricultural-bank-of-china-teams-up-with-state-owned-supply-chain-finance-platform-on-fintech-innovations/

China Banking News. "China Everbright Bank launches supply chain finance product using Ant Financial's blockchain platform", 18 March 2020. http://www.chinabankingnews.com/2020/03/18/china-everbright-bank-launches-supply-chain-finance-product-using-ant-financials-blockchain-platform/

Fadilpašić, Sead. "Are Chinese bitcoin mining farms moving to North America and why?" Crypto News, 12 February 2020. https://cryptonews.com/news/are-chinese-bitcoin-mining-farms-moving-to-north-america-and-5766.htm

Faridi, Omar. "Blockchain adoption: Nearly all of China's largest financial institutions are using distributed ledger technology", Crowdfundinsider, 23 April 2020. https://www.crowdfundinsider.com/2020/04/160515-blockchain-adoption-nearly-all-of-chinas-largest-financial-institutions-are-using-distributed-ledger-technology/

Financial Planning. "Asset managers have ESG, AI and ETFs top-of-mind for 2020", Financial Planning, 10 December 2019. https://www.financial-planning.com/news/asset-managers-have-esg-ai-and-etfs-top-of-mind-for-2020

Huang, Zheping. "China, home to the world's biggest cryptocurrency mining farms, now wants to ban them completely", SCMP, 9 April 2019. https://www.scmp.com/tech/policy/article/3005334/china-home-worlds-biggest-cryptocurrency-mining-farms-now-wants-ban

IMA Asia. "How blockchain can boost China's supply chain", IM Asia, March 2019. https://www.imaasia.com/blockchain-boosts-china-supply-chain/

Kong, Shuyao. "China's pending bitcoin 'mining catastrophe'", Decrypt, 8 March 2020. https://decrypt.co/21710/chinas-pending-bitcoin-mining-catastrophe

Lee, Amanda. "China's scandal-plagued P2P sector faces 'continued pressure' in 2020 amid tightening regulation", SCMP, 7 January 2020. https://finance.yahoo.com/news/chinas-scandal-plagued-p2p-sector-093000297.html

Li, Jane. "'Nude selfies for loans' scandal sheds light on China's rampant underground banking", SCMP, 7 December 2016. https://www.scmp.com/news/china/policies-politics/article/2052498/dozens-more-cases-revealed-chinese-students-forced-give

Microsoft Asia Writer. "FSI artificial intelligence + finance = intelligent investment", 24 September 2019. https://news.microsoft.com/apac/features/artificial-intelligence-finance-intelligent-investment/

Musharraf, Mohammad. "China's Internet Finance Association says blockchain is maturing", Cointelegraph, 15 April 2020. https://cointelegraph.com/news/chinas-internet-finance-association-says-blockchain-is-maturing

Pan, David. "From banking giants to tech darlings, China reveals over 500 enterprise blockchain projects", CoinDesk, 28 October 2019. https://www.coindesk.com/from-banking-giants-to-tech-darlings-china-reveals-over-500-enterprise-blockchain-projects

Partz, Helen. "China Everbright Bank uses Ant Financial's DLT for supply chain finance", Cointelegraph, 18 March 2020. https://cointelegraph.com/news/china-everbright-bank-uses-ant-financials-dlt-for-supply-chain-finance

Platonov, Ivan. "Five years of blockchain in China", EqualOcean, 4 March 2020. https://equalocean.com/high-tech/20200304-five-years-of-blockchain-in-china

Tatlow, Didi Kirsten. "To secure loans, Chinese women supply perilous collateral: Nude photos", New York Times, 16 June 2016. https://cn.nytimes.com/china/20160616/to-secure-loans-chinese-women-supply-perilous-collateral-nude-photos/en-us/

Tong, Qian, and Isabelle Li. "Tencent gives ground to UnionPay as Central Bank pushes for mobile payment integration", Caixin, 7 January 2020. https://www.caixinglobal.com/2020-01-07/exclusive-tencent-gives-ground-to-unionpay-as-central-bank-pushes-for-mobile-payment-integration-101501887.html

White, Maddy. "Standard Chartered invests in blockchain supply chain solution, eyes deep-tier SCF opportunities in China", Global Trade Review, 20 January 2020. https://www.gtreview.com/news/fintech/standard-chartered-invests-in-blockchain-supply-chain-solution-eyes-deep-tier-scf-opportunities-in-china/

Wood, Miranda. "China Construction Bank launches updated trade finance blockchain", Ledger Insights, October 2019. https://www.ledgerinsights.com/china-construction-bank-trade-finance-blockchain/

Chapter 6

Bork, Henrik. "The rise and all of Chinese bike sharing startups", Think:Act Magazine, 8 August 2019. https://www.rolandberger.com/en/Point-of-View/The-rise-and-fall-of-Chinese-bike-sharing-startups.html

Cadell, Cate. "As China's Didi looks abroad, challenges spring up at home", Reuters, 24 January 2018. https://www.reuters.com/article/us-china-didichuxing-analysis/as-chinas-didi-looks-abroad-challenges-spring-up-at-home-idUSKBN1FD117

Cheng, Si. "Sharing economy prompts industrial worker shortfall", China Daily, 1 May 2019. http://www.chinadaily.com.cn/a/201905/01/WS5c-c8d362a3104842260b96eb.html

Deng, Iris. "Alibaba merges food delivery units Ele.me and Koubei amid price war with Meituan in China's on-demand market", SCMP, 2 October 2018. https://www.scmp.com/tech/big-tech/article/2168265/alibaba-merges-food-delivery-units-eleme-and-koubei-amid-price-war

Knotts, Joey. "Meituan Bike inches up prices as Didi's Alternative stays steady", The Beijinger, 11 October 2019. https://www.thebeijinger.com/blog/2019/10/11/meituan-bike-inches-prices-didis-alternative-stays-steady

Liao, Rita. "Food delivered to the doorstep is not so cheap in China anymore", TechCrunch, 6 March 2019. https://techcrunch.com/2019/03/06/no-more-cheap-food-delivery-in-china/amp/

Liao, Rita. "Hellobike, survivor of China's bike-sharing craze, goes electric", TechCrunch, 15 July 2019. https://techcrunch.com/2019/07/14/china-micromobility-hellobike/

Luo, Weiteng. "A 'second spring' in store for shared two-wheelers?" China Daily, March 20, 2020. https://www.chinadailyhk.com/article/124959

Ma, Winston. "Here are 4 major bike-sharing trends from China after lockdown", World Economic Forum, 22 July 2020. https://www.weforum.org/agenda/2020/07/4-big-bike-sharing-trends-from-china-that-could-outlast-covid-19

Ouyang, Shaoxia. "Didi Qingju bicycles received over US\$1 billion in financing, bike sharing war is not ended", Technology Info Net, 17 April 2020.https://technology-info.net/index.php/2020/04/17/didi-qingju-bicycles-received-over-us-1-billion-in-financing-bike-sharing-war-is-not-ended/

Ricker, Thomas, and Andrew J. Hawkins. "Cities are transforming as electric bike sales skyrocket", The Verge, 14 May 2020. https://www.theverge.com/platform/amp/2020/5/14/21258412/city-bike-lanes-open-streets-ebike-sales-bicyclist-pedestrian

Russell, Jon. "Temasek jumps into China's bike-rental startup war with investment in Mobike", TechCrunch, 20 February 2017. https://techcrunch.com/2017/02/20/temasek-jumps-into-chinas-bike-rental-startup-war-with-investment-in-mobike/

Sen, Anirban, and Jane Lanhee Lee, "China's 'big gamble': Lessons from the bike sharing bust may hang over its A.I. boom", Fortune, 7 December 2019. https://fortune.com/2019/12/06/china-bike-sharing-investors-ai/

Sun, Yilei, and Brenda Goh. "China's Didi launches delivery service after coronavirus hammers ride-hailing demand", CNBC 16 March 2020. https://www.cnbc.com/2020/03/16/reuters-america-chinas-didi-launches-delivery-service-after-coronavirus-hammers-ride-hailing-demand.html (Reporting by

Wang, Jing. "A look at China's booming local-services market", Alizila, 24 March 2019. https://www.alizila.com/a-look-at-china-booming-local-services-market/

Chapter 7

Bernstein, Drew. "Despite risks, Chinese companies continue to flock to US exchanges", Marcum BP, 6 July 2020. https://crm.marcumbp.com/china-accounting-insights/despite-risks-chinese-companies-continue-to-flock-to-u.s.-exchanges

Brown, Eliot. "How Adam Neumann's over-the-top built WeWork: This is not the way everybody behaves", Wall Street Journal, 19 September 2019. https://www.wsj.com/articles/this-is-not-the-way-everybody-behaves-how-adam-neumanns-over-the-top-style-built-wework-11568823827

CCTV. "China securities regulator stresses dialogue with U.S. for achieving win–win results", 8 August 2020. https://english.cctv.com/2020/08/09/ARTILXplipw91S4dDm0qCEpK200809.shtml

Curran, Kevin. "Starbucks plans a big digital future in China, with Alibaba's help", TheStreet, Inc., 23 August 2018. https://realmoney.thestreet.com/articles/08/23/2018/starbucks-plans-big-digital-future-china-alibabas-help1

Economist. "The wave of unicorn IPOs reveals Silicon Valley's groupthink", 17 April 2019. https://www.economist.com/briefing/2019/04/17/the-wave-of-unicorn-ipos-reveals-silicon-valleys-groupthink

Economist. "Life is getting harder for foreign VCs in China", 9 January 2020. https://www.economist.com/business/2020/01/09/life-is-getting-harder-for-foreign-vcs-in-china

Ellis, Jack. "Starbucks enters China's agrifood investment scene with Sequoia Capital partnership", AFN, April 29, 2020. https://agfundernews.com/starbucks-enters-chinas-agrifood-investment-scene-with-sequoia-capital-partnership.html

Feng, Coco. "China's AI start-ups are closing more funding deals, yet they're still attracting less money than the US", SCMP, 22 January 2020. https://www.scmp.com/print/tech/venture-capital/article/3047161/chinas-ai-start-ups-are-closing-more-funding-deals-yet-theyre

Fitzgerald, Drew, and Sarah Krouse. "White House considers broad federal intervention to secure 5G future", Wall Street Journal, 26 June 2020. https://www.wsj.com/articles/white-house-federal-intervention-5g-huawei-china-nokia-trump-cisco-11593099054?mod=searchresults&page=1&pos=2

Jing, Meng. "Tencent partners with Starbucks to launch WeChat 'social gifting' feature", SCMP, 10 February 2017. https://www.scmp.com/tech/china-tech/article/2069850/tencent-partners-starbucks-launch-wechat-social-gifting-feature

Johnson, Khari. "CB Insights: AI startup funding hit new high of $26.6 billion in 2019", VentureBeat, 22 January 2020. https://venturebeat.com/2020/01/22/cb-insights-ai-startup-funding-hit-new-high-of-26-6-billion-in-2019/

Long, Danielle. "Starbucks partners with WeChat to launch social gifting feature for customers", The Drum, 14 February 2017. https://www.thedrum.com/news/2017/02/14/starbucks-partners-with-wechat-launch-social-gifting-feature-customers

Marvin, Rob. "2019: A no good, very bad year for unicorn tech companies", PCMag, 18 November 2019, https://www.pcmag.com/news/2019-a-no-good-very-bad-year-for-unicorn-tech-companies

Mohamed, Theron. "Softbank WeWork valuation 5 billion staggering drop", Business Insider, 8 November 2019. https://markets.businessinsider.com/news/stocks/softbank-wework-valuation-5-billion-staggering-drop-2019-11-1028673855

Pamuk, Humeyra. "Pompeo calls Nasdaq's strict rules a model to guard against fraudulent Chinese companies", Reuters, 4 June 2020. https://www.reuters.com/article/us-usa-china-pompeo-stocks-exclusive-idUSKBN23B2UC

Powell, Jamie, "NIO's New Year's Day surprise", Financial Times, 7 January 2020. https://ftalphaville.ft.com/2020/01/06/1578327570000/Nio-s-New-Years-Day-surprise/

Reuters. "WeWork debacle has unicorn investors seeking cover", 13 January 2020. https://www.reuters.com/article/us-funding-unicorns-analysis/wework-debacle-has-unicorn-investors-seeking-cover-idUSKBN1ZC0ZK

Rundell, Sarah. "Temasek seeks tomorrow's champions", Top100Funds.com, 12 October 2017. https://www.top1000funds.com/2017/10/temasek-seeks-tomorrows-champions/

Soo, Zen. "Excerpts of Q&A with Khazanah Nasional managing director", SCMP, 8 June 2017. https://www.scmp.com/tech/leaders-founders/article/2097400/excerpts-qa-khazanah-nasionalmanaging-director

Starbucks press release. "First-of-its-kind Starbucks virtual store to advance customer digital experiences in unimaginable ways in China", 13 December 2018. https://stories.starbucks.com/press/2018/starbucks-virtual-store-to-advance-customer-digital-experiences-in-china/

Steinberg, Julie, and Jing Yang. "Hong Kong's SenseTime considers $1 billion capital raise", Wall Street Journal, May 15, 2020. https://www.wsj.com/articles/hong-kongs-sensetime-considers-1-billion-capital-raise-11589561280

Swan, Jonathan, and Bethany Allen-Ebrahimian. "White House warns railroad retirement fund about Chinese investments", Axios, 8 July 2020. https://www.axios.com/white-house-railroad-retirement-fund-china-63c78920-6a8b-46d6-be0a-2b4db3927c05.html

Thorne, James. "China's VC industry bounces back after coronavirus-induced winter", Pitchbook, 2 April 2020. https://pitchbook.com/news/articles/chinas-vc-industry-bounces-back-after-coronavirus-induced-winter

Toh, Michelle. "'China's Netflix' is being investigated by the SEC for alleged fraud", CNN Business, 14 August 2020. https://www.cnn.com/2020/08/14/tech/iqiyi-baidu-sec-investigation-intl-hnk/index.html

Winck, Ben. "The IPO market is rebelling against many of 2019's money-losing unicorns", Business Insider, 2 October 2019. https://markets.businessinsider.com/news/stocks/ipo-market-outlook-trends-why-investors-rebelling-against-unicorns-implications-2019-9-1028570687#reason-1-weak-margins1

Yang, Jing. "Ernst & Young says it isn't responsible for Luckin Coffee's accounting misconduct", Wall Street Journal, 16 July 2020. https://www.wsj.com/articles/ernst-young-says-it-isnt-responsible-for-luckin-coffees-accounting-misconduct-11594909084

Yang, Jing. "'The gold standard': Why Chinese startups still flock to the U.S. for IPOs", Wall Street Journal, 13 August 2020. https://www.wsj.com/articles/the-gold-standard-why-chinese-startups-still-flock-to-the-u-s-for-ipos-11597313278?mod=hp_lead_pos3

Yang, Jing, and Juliet Chung. "Coffee's for closers: How a short seller's warning helped take down Luckin Coffee", Wall Street Journal, 29 June 2020. https://www.wsj.com/articles/coffees-for-closers-how-a-short-sellers-warning-helped-take-down-luckin-coffee-11593423002

Chapter 8

Borak, Masha. "Fake it till you make it: Fake followers boost Chinese celebs", Abacus News, 14 June 2019. https://www.abacusnews.com/digital-life/fake-it-till-you-make-it-fake-followers-boost-chinese-celebs/article/3014536

Boyd, Clark. "Data labeling is China's secret weapon in the connected car battle", Towards Data Science, 12 January 2020. https://towardsdatascience.com/data-labeling-is-chinas-secret-weapon-in-the-connected-car-battle-e8e395965380

CGTN Global Business. "Regulation on data and privacy will guarantee healthy dev't of China's digital economy", China Global Television Network, 14 May 2020. https://news.cgtn.com/news/2020-05-14/Regulation-on-data-and-privacy-benefits-China-s-digital-economy-QuetWXC42Q/index.html

Chen, Celia, and Tracy Qu. "As facial recognition tech races ahead of regulation, Chinese residents grow nervous about data privacy", SCMP, 27 November 2019. https://scmp.com/tech/policy/article/3039383/facial-recognition-tech-races-ahead-regulation-chinese-residents-grow

Deng, Iris. "Didi Chuxing boss Jean Liu says car-pooling controversy was bigger 'blow' than her own fight against cancer", SCMP, 28 November 2019. https://scmp.com/tech/start-ups/article/3039571/didi-chuxing-boss-jean-liu-says-car-pooling-controversy-was-bigger

Gan, Nectar, and David Culver. "CNN China is fighting the coronavirus with a digital QR code. Here's how it works", CNN, April 16, 2020. https://www.cnn.com/2020/04/15/asia/china-coronavirus-qr-code-intl-hnk/index.html

Hart, Robert. "Chinese data protection: The move towards a comprehensive law", Global Data Review, 14 July 2020. https://globaldatareview.com/data-privacy/chinese-data-protection-the-move-towards-comprehensive-law

Japan Times. "Green or red light: China coronavirus app is ticket to everywhere", 13 May 2020. https://www.japantimes.co.jp/news/2020/05/13/asia-pacific/china-coronavirus-app/#.Xt72JTpKhPY

Jing, Yu. "Facial recognition, deep fake: What China's Civil Code means for privacy protection", China Global Television Network, 26 May 2020. https://news.cgtn.com/news/2020-05-26/What-China-s-Civil-Code-means-for-privacy-protection-QNOwx9K5Og/index.html

Ma, Winston. "China is waking up to data protection and privacy", World Economic Forum, November 2019. https://www.weforum.org/agenda/2019/11/china-data-privacy-laws-guideline/

Ma, Winston. "15 years on, what challenges does Facebook face today", China Global Television Network, 4 February 2019. https://news.cgtn.com/news/3d3d514f3163444e32457a6333566d54/index.html

Ouyang, Shaoxia. "The driver's sexual assault live broadcast, Didi spent a moment in shock", Technology Info Net, 12 June 2020.https://technology-info.net/index.php/2020/04/17/didi-qingju-bicycles-received-over-us-1-billion-in-financing-bike-sharing-war-is-not-ended/

Ouyang, Shijia, and Chen Jia. "New guideline to better allocate production factors", China Daily, 10 April 2020. https://www.chinadaily.com.cn/a/202004/10/WS5e903fd7a3105d50a3d15620.html

Pan, Che. "China pushes nationwide buildout of new 'Big Data' centres to speed up digital transformation of industries", SCMP, 15 May 2020. https://www.scmp.com/tech/policy/article/3084630/china-pushes-nationwide-buildout-new-big-data-centres-speed-digital

SCMP. "Think China's data is an unbeatable AI advantage? A new report says otherwise", 18 July 2019. https://scmp.com/news/china/society/article/3019067/think-chinas-data-unbeatable-ai-advantage-new-report-thinks

Singer, Natasha, and Choe Sang-Hun. "As coronavirus surveillance escalates, personal privacy plummets", New York Times, 23 March 2020. https://www.nytimes.com/2020/03/23/technology/coronavirus-surveillance-tracking-privacy.html?referringSource=articleShare

Wakabayashi, Daisuke, Jack Nicas, Steve Lohr, and Mike Isaac. "Big tech could emerge from coronavirus crisis stronger than ever", New York Times, March 23, 2020. https://www.nytimes.com/2020/03/23/technology/coronavirus-facebook-amazon-youtube.html

Xinhua News. "China focus: Data-labeling: the human power behind artificial intelligence", 17 January 2019. http://www.xinhuanet.com/english/2019-01/17/c_137752154.htm

Xu, Tony. "Douyin campaign bans 2 million accounts for faking fans", Technode, Jan 9, 2020. https://technode.com/2020/01/09/douyin-campaign-bans-2-million-accounts-for-faking-fans/

Zhao, Yusha. "Beijing busts app for inflating likes, shares of pop idol", Global Times, 11 June 2019. http://www.globaltimes.cn/content/1153860.shtml

Chapter 9

Bloomberg. "China AI champions like SenseTime are scrambling to survive Trump's blacklist", SCMP, 29 October 2019. https://scmp.com/tech/start-ups/article/3035279/china-ai-champions-sensetime-are-scrambling-survive-trumps-blacklist

Clark, Harry, and Betty Louie. "United States: Grindr and PatientsLikeMe outcomes show non-cleared transactions' exposure to CFIUS scrutiny, especially when PII is involved", Mondaq, 8 May 2019. https://www.mondaq.com/unitedstates/Government-Public-Sector/804096/Grindr-And-PatientsLikeMe-Outcomes-Show-Non-Cleared-Transactions39-Exposure-To-CFIUS-Scrutiny-Especially-When-PII-Is-Involved

CSIS. "RESOLVED: Japan could lead global efforts on data governance", 27 June 2019. https://www.csis.org/analysis/resolved-japan-could-lead-global-efforts-data-governance

Disis, Jill. "CNN: A new world war over technology", CNN Business, 10 July 2020. https://www.cnn.com/2020/07/10/tech/us-china-global-tech-war-intl-hnk/index.html

Economist. "ByteDance is going from strength to strength – America doesn't like it one bit", 18 April 2020. https://www.economist.com/business/2020/04/18/bytedance-is-going-from-strength-to-strength

Ferek, Katy Stech. "National security regulator to take closer look at privacy risks in foreign investors' U.S. deals", Wall Street Journal, 13 February 2020. https://www.wsj.com/articles/national-security-regulator-to-take-closer-look-at-privacy-risks-in-foreign-investors-u-s-deals-11581600600

Jing, Meng. "Your money's not wanted HERE: Chinese-led bid for stake in high-res map maker rejected by US", SCMP, 27 September 2017. https://scmp.com/business/companies/article/2113119/your-moneys-not-wanted-here-chinese-led-bid-stake-high-res-map

Khan, Mehren. "EU floats plan for €100 billion sovereign wealth fund", Financial Times, 23 August 2019. https://www.ft.com/content/033057a2-c504-11e9-a8e9-296ca66511c9

Japan Times. "Japan display bailout plan gets shareholder approval", 26 March 2020. https://www.japantimes.co.jp/news/2020/03/26/business/corporate-business/japan-display-bailout/

Li, Tao. "How China's 'Big Fund' is helping the country catch up in the global semiconductor race", South China Morning Post, 10 May 2018. https://scmp.com/tech/enterprises/article/2145422/how-chinas-big-fund-helping-country-catch-global-semiconductor-race

Li, Xiang. "PE firm Canyon Bridge shifting focus to Europe", China Daily, 9 February 2018. https://www.chinadaily.com.cn/a/201802/09/WS5a-7d0205a3106e7dcc13bbf1.html

Ma, Winston. "Can GDD create a link in cyberspace", China Global Television Network, 10 September 2019. https://news.cgtn.com/news/2019-09-10/Can-GDD-create-a-link-in-cyberspace–JSwaelOhXO/index.html?from=singlemessage&isappinstalled=0

Mercury News. "Fairchild turns down Chinese offer", 17 February 2016. https://www.mercurynews.com/2016/02/17/fairchild-turns-down-chinese-offer/

Musgrave, Paul. "The slip that revealed the real Trump doctrine", Foreign Policy, 2 May 2019. https://foreignpolicy.com/2019/05/02/the-slip-that-revealed-the-real-trump-doctrine/

Newton, Casey. "How Grindr became a national security issue", The Verge, 28 March 2019. https://www.theverge.com/interface/2019/3/28/18285274/grindr-national-security-cfius-china-kunlun-military

Sanger, David. "Japanese purchase of chip maker cancelled after objections in US", New York Times, 17 March 1987. https://www.nytimes.com/1987/03/17/business/japanese-purchase-of-chip-maker-canceled-after-objections-in-us.html

Soo, Zen, Sarah Dai, and Meng Jing. "Lagging in semiconductors, China sees a chance to overtake the US with AI chips as 5G ushers in new era", SCMP,18 September 2019. https://scmp.com/tech/enterprises/article/3027775/lagging-semiconductors-china-sees-chance-overtake-us-ai-chips-5g

Xiang, Li. "PE firm Canyon Bridge shifting focus to Europe", China Daily, 9 February 2018. https://www.chinadaily.com.cn/a/201802/09/WS5a7d0205a3106e7dcc13bbf1.html

Yu, Yihan. "ByteDance names ex-Disney executive as COO and TikTok CEO", Nikkei Asian Review, 19 May 2020. https://asia.nikkei.com/ Business/Technology/ByteDance-names-ex-Disney-executive-as-COO-and-TikTok-CEO

Zhong, Raymond. "U.S. blocks a Chinese deal amid rising tensions over technology," New York Times, 23 February 2018. https://www.nytimes. com/2018/02/23/technology/china-microchips-cfius-xcerra.html

Chapter 10

African Union. "The Draft Digital Transformation Strategy for Africa (2020–2030)", tralac. https://www.tralac.org/documents/resources/ african-union/3013-the-draft-digital-transformation-strategy-for-africa-2020-2030/file.html

Agence France-Press. "Germany aims to shield tech firms from foreign takeovers", SCMP, 28 November 2019. https://scmp.com/tech/policy/ article/3039814/germany-aims-shield-tech-firms-foreign-takeovers

Bright, Jake. "Driving deep into Africa's blossoming tech scene", Tech-Crunch, 31 May 2019. https://techcrunch.com/2019/05/31/diving-deep-into-africas-blossoming-tech-scene/ accessed on Jan. 19, 2020

Davis, River. "Japan plans national champion to challenge Huawei", Wall Street Journal, 25 June 2020, https://www.wsj.com/articles/japan-plans-national-champion-to-challenge-huawei-11593091392

Dickman, Steve. "US crackdown on foreign biotech investment makes us poorer, not safer", Forbes, 24 May 2019. https://www.forbes.com/ sites/stevedickman/2019/05/24/us-crackdown-on-foreign-biotech-investment-makes-us-poorer-not-safer/#311dbd558197

Disruptive Tech ASEAN. "Malaysia gets first global data center, thanks to Alibaba Cloud", 31 October 2017. https://disruptivetechasean. com/big_news/malaysia-gets-first-global-data-center-thanks-to-alibaba-cloud/

Economist. "A global timepass economy: How the pursuit of leisure drives internet use—Movies, not grain prices, are bringing the poor world online", 8 June 2019. https://www.economist.com/briefing/ 2019/06/08/how-the-pursuit-of-leisure-drives-internet-use

Gal, Danit. "The U.S.–China–Israel technology triangle", Council on Foreign Relations, 30 July 2019. https://www.cfr.org/blog/us-china-israel-technology-triangle

Hartcher, Peter. "Revealed: Why the sale of Ausgrid to Chinese buyers was vetoed", Sydney Morning Herald, 28 May 2018. https://www.smh. com.au/opinion/revealed-why-the-sale-of-ausgrid-to-chinese-buyers-was-vetoed-20180528-p4zhxh.html

Henderson, Richard. "Australia adds national security test to foreign investment rules", Financial Times, 5 June 2020. https://www.ft.com/content/f3a1833a-493a-43a8-babb-f2ff3dbc4cc1

Kamiya, Marco, and Winston Ma, "Sovereign investment funds could be the answer to the SDGs", World Economic Forum, December 2019. https://www.weforum.org/agenda/2019/12/sovereign-wealth-funds-sdgs/

Ma, Winston. "The Digital Silk Road brings the new momentum of green transformation for the BRI", World Economic Forum, September 2019. https://www.weforum.org/agenda/2018/09/could-a-digital-silk-road-solve-the-belt-and-roads-sustainability-problem/

Ma, Winston. "The Digital Silk Road for the next billion users", Milken Institute Review, 2018. http://www.milkenreview.org/articles/the-digital-silk-road-for-the-next-billion-users

New York Times. "Really? Is the White House proposing to buy Ericsson or Nokia?" 13 February 2020. https://www.nytimes.com/2020/02/07/business/dealbook/bill-barr-huawei-nokia-ericsson.html

The Nation. "Alibaba's Taobao Global kicks off world initiative in Thailand", 8 January 2020. https://www.nationthailand.com/business/30380293

Pan, David. "Meet Red Date, the little-known tech firm behind China's big blockchain vision", CoinDesk, 24 April 2020. https://www.coindesk.com/meet-red-date-the-little-known-tech-firm-behind-chinas-big-blockchain-vision

Pop, Valentina. "EU moves to shrink Chinese, U.S. influence in its economy", Wall Street Journal, 17 June 2020. https://www.wsj.com/articles/eu-moves-to-shrink-chinese-u-s-influence-in-its-economy-11592393124?mod=searchresults&page=1&pos=7

Pop, Valentina. "Protectionism spreads globally with the new coronavirus", Wall Street Journal, 30 May 2020. https://www.wsj.com/articles/protectionism-spreads-globally-with-the-new-coronavirus-11590779442?mod=searchresults&page=1&pos=3

Satariano, Adam, and Monika Pronczuk. "Europe, overrun by foreign tech giants wants to grow its own", New York Times, 19 February 2020. https://www.nytimes.com/2020/02/19/business/europe-digital-economy.html

Scott, Jason. "Australia to toughen foreign investment laws amid China spat", Bloomberg, 4 June 2020. https://www.bloomberg.com/news/articles/2020-06-04/australia-to-boost-foreign-investment-rules-for-sensitive-assets

Scheck, Justin, Rory Jones, and Summer Said. "A prince's $500 billion desert dream: Flying cars, robot dinosaurs and a giant artificial moon", Wall Street Journal, 25 July 2019. https://www.wsj.com/articles/a-princes-500-billion-desert-dream-flying-cars-robot-dinosaurs-and-a-giant-artificial-moon-11564097568

Singh, Karamjit. "Alibaba Cloud doubles down on ASEAN region", Digital News Asia, October 1, 2019. https://www.digitalnewsasia.com/business/alibaba-cloud-doubles-down-asean-region

Sung, Michael. "China's national blockchain will change the world", CoinDesk, 24 April 2020. https://www.coindesk.com/chinas-national-blockchain-will-change-the-world

Tomás, Juan Pedro. "US lawmakers propose $1 billion fund to replace Chinese network gear", RCR Wireless News, 26 September 2019. https://www.rcrwireless.com/20190926/5g/us-lawmakers-propose-1-billion-fund-replace-huawei-gear

United Nations Conference on Trade and Development. "UNCTAD Digital Economy Report 2019", United Nations Conference on Trade and Development, 4 September 2019. https://unctad.org/en/PublicationsLibrary/der2019_en.pdf

Wang, Jing. "Taobao Villages driving 'inclusive growth' in rural China", Alizila, 25 November 2019. https://www.alizila.com/taobao-villages-driving-inclusive-growth-rural-china/

Xinhua News Agency. "Spotlight: Southeast Asian e-commerce boosted by Chinese expertise", 2 November 2019. http://www.xinhuanet.com/english/2019-11/02/c_138522877.htm

China's Digital Innovation Challenges US Tech Supremacy

"As the United States and China head down the ill-conceived path of digital and technology decoupling, *The Digital War* must be read before these policies become irreversible. Winston Ma's insider's view of China's digital development takes the reader down a path too rarely tread by those trying to understand China".

– Stephen A. Orlins,
President, National Committee on US–China Relations

"If you're not paying attention to China's digital transformation, you're missing the emergence of the world's new technology superpower, one whose innovations will drive rapid economic advances and alter the balance of global economic power. Thankfully, Winston Ma has delivered exactly the kind of comprehensive, timely, and thought-provoking account you need to get up to date. Don't miss this vitally important read".

– Michael J. Casey, Co-author,
The Truth Machine: The Blockchain and the Future of Everything

"The Fourth Industrial Revolution may impose significant challenges for many of the smaller countries, especially those that are lacking technology resources and have large labor forces that might be replaced by artificial intelligence (AI). But it may also be a leap-frogging opportunity for those countries, if they embrace the new technology revolution just as keenly, and the Digital Silk Road from China provides the necessary support. Winston Ma's latest book on the digital economy is a cutting-edge text that spotlights the digital transformation in China and its global impact through the Silk Road in cyberspace".

– Professor Klaus Schwab,
Founder and Executive Chairman of the World Economic Forum

"China is determined to make innovation an engine for the next stage of the country's development, and no sector has been more creative or dynamic than the mobile economy, which in some areas, has surpassed even the United States. Winston Ma's deep dive into this fiercely competitive, constantly evolving industry dissects the companies, personalities, and forces that are transforming China and that inevitably will influence commerce far beyond its shores".

– John L. Thornton,
Co-chairman, Brookings Institution

"Having worked at both, the Wall Street and China Investment Corporation, Winston Ma seems eminently qualified to give both Western and Chinese readers a unique perspective on the development of China's digital economy. This is the go-to-guide for companies and investors seeking to understand China's grand tech ambitions, what the Chinese players' strategies are, and where the huge opportunities are".

– Dr. Chen Datong,
Founding Partner, West Summit Capital

Winston Ma is an investor, attorney, author, and adjunct professor in the global digital economy. Most recently, for 10 years, he was Managing Director and Head of the North America Office for China Investment Corporation (CIC), China's sovereign wealth fund. Prior to that position, Ma served as the deputy head of equity capital markets at Barclays Capital, a vice president at J.P. Morgan investment banking, and a corporate lawyer at Davis Polk & Wardwell LLP. He is the author of best-selling books, including: *Investing in China: New Opportunities in a Transforming Stock Market, China's Mobile Economy: Opportunities in the Largest and Fastest Information Consumption Boom,* and *The Hunt for Unicorns: How Sovereign Funds Are Reshaping Investment in the Digital Economy.* Ma was selected a 2013 Young Global Leader at the World Economic Forum, and in 2014, he received the New York University Distinguished Alumni Award.

Index

357